THE MIDWEST
FARMER'S
DAUGHTER

D1254152

THE MIDWEST FARMER'S DAUGHTER

IN SEARCH OF AN AMERICAN ICON

BY ZACHARY MICHAEL JACK

LIBRARY
FRANKLIN PIERCE UNIVERSIT
RINDGE NH 03461

PURDUE UNIVERSITY PRESS — WEST LAFAYETTE, INDIANA

Copyright 2012 by Purdue University. All rights reserved.

Printed in the United States of America.

Library of Congress Cataloging-in-Publication Data
Library of Congress Cataloging-in-Publication Data

Jack, Zachary Michael, 1973-
  The Midwest farmer's daughter : in search of an American icon /
Zachary Michael Jack.
      p. cm.
  Includes bibliographical references.
  ISBN 978-1-55753-619-8 (paper : alk. paper) - ISBN 978-1-61249-219-3
(ePDF) - ISBN 978-1-61249-218-6 (ePUB) 1. Women farmers-Middle
West. 2. Farm life-Middle West. 3. Middle West-Social life and customs.
4. Women farmers-Middle West-Public opinion 5. Farm life-Middle
West-Public opinion 6. Middle West-Public opinion 7. Women in mass
media. 8. Farm life in mass media. 9. Popular culture-United States. 10.
Public opinion-United States. I. Title.
  S521.5.M53J33 2012
  630.82-dc23
                                        2012004918

Cover image: *The Prairie Was Her Playground*
copyright 2008 Lara Blair Images / www.modernprairiegirl.com

TO THE DAUGHTERS
THAT MADE THE SEEDBED
GAIL
BARBARA
SUSAN
AND PATRICIA
AND IN MEMORY
OF THOSE
THAT PLOWED THE FURROW
AMBER JANE PICKERT
AND JULIA MAE PUFFER

Three generations of Midwest farmers' daughters—the author's great-great-great grandmother, great-great grandmother, and great-grandmother holding baby Edward Lee Jack on the Jack family farm around 1920.

CONTENTS

Preface: Pioneering Women   ix

PART ONE

CHAPTER ONE
The Gingham Girl in the Google Age   3

CHAPTER TWO
The Midwest Farmer's Daughter   25

CHAPTER THREE
How Ya Gonna Keep 'Em Down on the Farm?   47

CHAPTER FOUR
Raising Farmer Jane   61

CHAPTER FIVE
The Chores of Being a Farm Girl   89

CHAPTER SIX
Welk Girls and Daisy Dukes   117

CHAPTER SEVEN
Milkmaids in Manhattan   133

PART TWO

CHAPTER EIGHT
Little Houses on the Prairie   151

CHAPTER NINE
Future Farm Daughters of America   163

CHAPTER TEN
Ag-vocating Women   187

CHAPTER ELEVEN
Community-Supported Agriculture   211

CHAPTER TWELVE
Female Farmers   231

CHAPTER THIRTEEN
Farmerettes in the Farm City   239

CHAPTER FOURTEEN
Her Daughter Has a Dynamo   255

Acknowledgments   259

Selected Bibliography   261

PREFACE

PIONEERING WOMEN

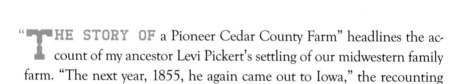

"THE STORY OF a Pioneer Cedar County Farm" headlines the account of my ancestor Levi Pickert's settling of our midwestern family farm. "The next year, 1855, he again came out to Iowa," the recounting goes, "bringing not only his wife and son, but his father and mother also."

So begins the history of my people, and the plot advanced as they planted seeds real and metaphorical in the good midwestern dirt. Yet while the presumed protagonist of the pioneering drama, Levi, earns multiple mentions, his helpmate in life remains nameless but for the unremarkable moniker "wife." As the pioneering history unfolds with its breathless stories of the "coldest winter in Iowa's history" when, "for forty consecutive days it did not thaw," climaxing in tales of hangings and horses thieves on a frontier where "trees in the vicinity [had] been decorated with the bodies of desperadoes," the very name of Levi's life partner is lost to the wind, a sound and fury, signifying nothing.

By my late twenties I could recite the names of my male farming fore-bears on the Pickert side as far back as the early 1800s. But had I been asked to name a wife or daughter predating my great-grandmother, I would surely have come up empty, not because I was a poor student of genealogy, but because the names were seldom found in print. It wasn't until my thirties, in fact, that a bundle of letters I'd chanced upon revealed to me the hearts of the family's farm daughters, making them, quite literally, something to have and to hold.

In one letter datelined 1867, Syracuse, New York, Eliza Smith writes her long-lost sister, my great-great-great-grandmother Sally Pickert, wife of Levi, to observe, "This world is full of sorrow and disappointment. We was very glad to hear from you once more and know that you are alive, but I think you will not live long if you keep on working so hard as you do. What profit will it be to you to have it said that you were rich after you are dead? I have seen the folly of working so hard for greater riches and see them take wings and fly away. Sally, you do not know how much I want to see you and talk with you. I have so much to say that I can't write . . ."

The earliest missive Sally saved, dated 1855, would have arrived while she, Levi, and her in-laws shared a one-room schoolhouse with four other families, according to a 1962 article in the *Cedar Rapids Gazette* entitled "Three Generations of Pickerts Have Lived in Mechanicsville House." The Pickerts arrived in Davenport, Iowa, in 1854 by train from Waterton, New York, the *Gazette's* Amber Jackson reported, and began walking west until they came upon the 200 acres of black earth that would become our Iowa Heritage Farm, purchasing the ground from David Platner for the bargain price of $10 an acre. Jackson's recounting makes no mention of the female partners in the enterprise beyond the wife-obscuring umbrella "The Levi Pickerts," nor does it mention the baby Sally lost in that first unforgiving year in the Heartland.

"Your Uncle Ben told me he would give $100 if you would come back," another New York relative named William Wallis conveyed in his own note to his far-flung Iowa relatives. But there would be no such turning back for pioneering families, not for love or money, no diminution of the arcadian dream of mother and father, daughter and son, cultivating the country-side for generations. To the yeoman the dream seemed unerring, the yield perennial. The farmer's son may have made the harvest possible, but the daughter made it worthwhile.

"**AND STILL TODAY** the same reverie comes each spring, the scenes are the same. Everything, in fact, save for the little girl," my great-grandfather Walter Thomas Jack wrote of his own farmer's daughter, Helen, in his book *The Furrow and Us*. "She has grown up now and has gone, but imagination keeps her on the set, and her role will always be that of the leading lady." Still, even as Grandpa Walt penned homage to his own long-gone girl of 1943, powerful cultural and economic changes had already swept many a farm's "leading lady" from her bucolic perch. Lovely Helen Jack proved a case in point: she had grown up, married, and moved off the farm into secretarial work at the John Deere Company in Moline, Illinois, leaving her father behind to pine. Great-grandpa's "reverie" of a girl "bedecked in Easter finery" was a "song," he insisted, that "clings to us throughout our whole lives." "Neither time nor eternity can take away the particularity, and as time passes, the charm of it remains radiant, immortal," he opined, though he might have classed the same phenomena as a bona fide haunt.

It had been some Rip Van Winkle sleep, surely, in which she had slipped away, his girl of spring. In the pages of Walt's wartime glossies, after all, agricultural daughters like Helen Jack weren't pictured at desks taking rote dictation for ungrateful bosses; instead they were shown astraddle tractors, tilling the good earth while wearing straw hats and white work shirts rolled at the sleeve, frisky farm collies following in their wake. The advertisements in great-grandpa's *Collier's* toasted a girl who had become a wartime rallying cry and a Madison Avenue ad man's dream, as enchanting to a nation of urbanites as to the farm mothers and fathers who kept the memory of her alive back home. "Uncle Sam's traveling men . . . more than 8,000,000 strong are on the march all over the world," read an ad for Easy washing machines. "But they're never out of touch with home. Thanks to the farmer's daughter, and the millions like her, our boys can count on a steady stream of the food and other supplies they need to win. Easy salutes the farmer's daughter . . . and all the other heroines fighting so valiantly on the homefront."

By 1949 the farm daughter had indeed emerged as a cultural leading lady, becoming the stuff of silver screens and doughboy centerfolds. Two years earlier a film entitled *The Farmer's Daughter* had been honored as *Life* magazine's movie of the week, and suddenly here she was—the rural equivalent of a real-life Gibson girl in the pages of America's most iconic lifestyle magazine—Virginia Jensen of Blair, Nebraska, reclining in a bed of hay while wearing a button-down blouse and short-shorts, a hayseed slipped provocatively between her teeth. Jensen, the caption announced, had been declared the "Ideal Farmer's Daughter" at a Chicago National Farm Show where contestants vying for the crown had made their costumes out of feedbags. In an ironic foreshadowing of the cultural postmortem to come, the vivacious, victorious Jensen earned as her prize tickets to *Death of a Salesman*.

The *Life* headline proclaiming Jensen's reign in December of 1949 read, "Yes, My Darling Daughter," but it might just as well have read, "Yes, Virginia, there is a Santa Claus." By 1950 the Midwest farmer's daughter had become part celebrity, part myth, part scapegoat, and part salvation. Soon enough she, like Santa, would qualify as a cultural apparition kept alive mostly as an article of faith.

What follows is the story of how the gingham girl became the glamour girl became a generation's most important ghost. Like all questions, the question of what became of the gingham girl, of women like my mother and grandmother and aunts who put the heart in Heartland, is born of mystery and longing and incompletion. Following it means coming boot-to-boot with a whodunit, toe-to-toe with a mystery regional, familial, and societal, economic and cultural, midwestern and American, yours and mine, one whose song may have ended, but whose melody lingers on.

PART ONE

"*The Country Girl and Her Pets. 'The quietness of the country permits a greater
spiritual and mental growth, with its abundance of life, plant and animal,
which challenges the mind to discover its secrets.'*"

CHAPTER ONE

THE GINGHAM GIRL IN THE GOOGLE AGE

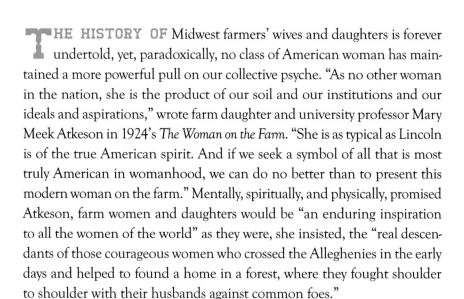

THE HISTORY OF Midwest farmers' wives and daughters is forever undertold, yet, paradoxically, no class of American woman has maintained a more powerful pull on our collective psyche. "As no other woman in the nation, she is the product of our soil and our institutions and our ideals and aspirations," wrote farm daughter and university professor Mary Meek Atkeson in 1924's *The Woman on the Farm*. "She is as typical as Lincoln is of the true American spirit. And if we seek a symbol of all that is most truly American in womanhood, we can do no better than to present this modern woman on the farm." Mentally, spiritually, and physically, promised Atkeson, farm women and daughters would be "an enduring inspiration to all the women of the world" as they were, she insisted, the "real descendants of those courageous women who crossed the Alleghenies in the early days and helped to found a home in a forest, where they fought shoulder to shoulder with their husbands against common foes."

At the time my ancestor Levi Pickert settled what is now our Iowa Heritage Farm, more than half of 26 million Americans made their living at farming. By 1915, the year the dean of women at Northwestern University, Martha Foote Crow, wrote the first true examination of the US farm daughter entitled *The American Country Girl*, among an estimated 98 million Americans—more than 6 million ages 15 to 29—could lay claim to the title. The number would soon reach 7 million, Crow promised, unless "these six millions shall have run away to build their homes and rear their children in the hot, stuffy, unsocialized atmosphere of the town, leaving the happy gardens without the joyous voices of children, the fields without sturdy boys to work them, the farm homes without capable young women to—shall I say, to *man* them?"

For a century and a half the Midwest country girl existed as something of an enigma, as fated and feted as a fairy. As Crow herself asked, "Where is the tall young daughter and where are the papers for her and the books about her needs? It seems that she has yet to find a voice." It was far easier, she observed, to plumb the psyche of the farmer's son than to understand the wants and whimsies of the farmer's daughter. "Since the days of Mother Eve the woman young and old has been adapting herself and readapting herself," the author lamented, "until, after all these centuries of constant practice, she has become a past master in the art of adaptation." She had, claimed Crow, been "consciously or unselfconsciously hiding herself." Still, when one lists the sweethearts of any American generation up to and including my own, generation X, one finds a farmer's daughter at or near the top. "There was the great romance of all America," Atkeson alleged of the country girl, "the woman in the sunbonnet, and not, after all, the hero with the rifle." Still, Atkeson echoed Crow's query nearly verbatim, begging the question, "Who has written her story? Who has painted her picture?"

In 1973, the forgettable year at the end of which I entered the world, the preternaturally buxom country girl of Sevierville, Tennessee, Dolly Parton released her hit chart-topper "Jolene," and farmer's and miner's daughter Loretta Lynn of Butcher Holler, Kentucky, had, not so long before, reached number one with her autobiographical anthem "Coal Miner's Daughter."

On television Donna Douglas's Elly May Clampett captured the fancy of audiences for nearly three hundred episodes of *The Beverly Hillbillies* from 1962 to 1971, mingling sex symbol with rural route ingénue, while the equally blond and beautiful Eva Gabor played a comely if not poorly suited farmer's wife on *Green Acres*. When *Green Acres* and *The Beverly Hillbillies* ended their storied runs in 1971, a new generation of fictional farmers' daughters stood ready to step from the blank fields of Los Angeles into the promising "Back 40" of some famed Hollywood studio lot. Six months after I drew first breath at Mercy Hospital in Iowa City, Iowa, the two-hour pilot for *Little House on the Prairie* debuted, featuring quintessential pioneer girl Laura Ingalls Wilder, whose timeless charms vaulted the show into the Top 10 Nielsen Ratings one season later. From 1979 onward Friday's nights before *Dallas* meant Catherine Bach's portrayal of farmer's daughter Daisy Duke in *The Dukes of Hazzard*, a show that likewise cracked the Top 10 and named a genre of tightly-fitting jean shorts.

The Midwest farmer's daughter wasn't hard to find in the era of Nixon, Ford, and Carter. She was my mother, and my aunts, brightening the farmsteads they shared with their husbands; she was my sister, Natasha; she was my sister's teacher and our farm neighbor, Dorothy Ferguson; she was our neighborhood girl heartthrobs, Amy Kohl and the Light twins, just up the road. We didn't know it at the time, but all of us were powerfully imprinting on an icon that had largely become, outside our home region, a fiction if not a phantasm. Indeed, by the time I entered kindergarten in Lisbon, Iowa, for the 1979–1980 school year, the famous Iowa's Favorite Farmer's Daughter contest sponsored by WMT radio in Cedar Rapids, Iowa, had ceased its decades-long run. And by the time the 1980s farm crisis beset the Midwest and found me staring down the barrel of middle school, poor Daisy Duke had become the last country girl standing on popular television. While on the show Uncle Jesse's farm was small potatoes compared to the 500 acres typical of midwestern spreads, for the overwhelming majority of 238 million American viewers, Jesse's daughter was as close to the genuine article as they were likely to find. Daisy Duke was the only vehicle by which an adult farmer's daughter, however sexist, might enter their living room.

PREPARATORY TO MY search for the whereabouts of an American icon, I entered *farmer's daughter* into dozens of search engines and databases. All yielded results predictably caricatured. I clipped an article in *Newsweek* featuring Winthrop, Iowa, farmer's daughter turned actress Michelle Monaghan. Monaghan had made news as the sultry costar opposite George Clooney in *Syriana* and Val Kilmer in *Kiss Kiss Bang Bang*, and the two-page vertical spread read, "But just because Monaghan is a genuine Iowa farm girl, it doesn't mean she's just some farm girl from Iowa. In next fall's comic noir *Kiss Kiss Bang Bang*, she'll get her first starring role, playing a fierce, frisky, struggling actress who's wrapped up in a murder. The role includes a brief nude scene and enough four-letter words to scorch a harvest." The same cheeky, celebrity gossip tone prevailed in an "Ask Marilyn" column clipped from *Parade* magazine in 2006, in which columnist Marilyn vos Savant asked her readers to address the ironical "what if" in "what if *Sex and the City*'s Sarah Jessica Parker were a farmer's daughter instead?" America's answers proved hokier still. "The scarecrow would wear Prada," one wit from Escondido, California, quipped, while another weighed in, "Her show would be called *Sex and the Silo*."

The farmer's daughter took another pop cultural hit in the urban press when R. Tripp Evans's biography of Grant Wood, who grew up just down the road from my own Iowa farm, appeared to confirm that Wood was, as one cad in the blogosphere put it, "gay as a purse-full of kittens." The revelation was hailed as a victory by the gay community, which could now reasonably claim the world's most recognizable painting, *American Gothic*, as a wryly ambiguous statement not just about farmer's daughters, but about repressed Middle American heterosexuality more generally. Though in a letter to the *Des Moines Register* Grant's sister Nan, the she-model for the famous canvas, had once characterized the painting's rural miss as "one of those terribly nice and proper" farmer's daughters, Evans's more fraught interpretation evidenced changing readings of the American country girl. The author hones in on the woman's loosened hair, in particular, calling it "a motif that traditionally signals a woman's abandonment of sexual inhibitions" and Wood's "connection to jokes about promiscuous farmer's daughters."

Yet for all this ironic attention, the country girl has enjoyed a renaissance at exactly the time when finding a real one, in the flesh, has become vexsome even in Grant Wood's Iowa, where nearly 65 percent of the population of the Hawkeye State is now classed as urban. Meanwhile, popular books like Jennifer Worick's *The Prairie Girl's Guide to Life* trade well and widely on a latter-day identification with farm daughters past. "Gals are taking up sewing, enrolling in meat-curing classes, checking out quilting and yarn expos, creating shrines to Martha Stewart. They may not realize it, but each of these women is a prairie gal. *You* are a prairie gal," Worick reminds eager readers. Geography, she insists, has nothing to with the country girl ethos: "I can live in the city, far away from my homestead, and feel that bond." Never mind that Martha Stewart was born in Jersey City, New Jersey, and that Stewart, unlike Laura Ingalls Wilder, was convicted of conspiracy, obstruction, and making false statements to federal investigators. Somehow Stewart had been welcomed into the increasingly symbolic fold of would-be "country girls," even though her attorneys protested her incarceration in a West Virginia prison on the grounds that it was too remote. "Prairie girl" Martha Stewart wanted civilized Connecticut or Florida, not the boonies, for her stay in the hoosegow.

Online a whole series of farm and prairie girl blogs cropped up with titles like *Confessions of a Farm Wife* featuring 30- and 40-somethings who had spent their formative years in the city only to return with much ado to a farm or acreage, or to conjure the rustic childhood they never had from the friendlier confines of the metropolis. "Oh, I am a city girl at heart," read the opening line of an August 2011 post on *Confessions of a Farm Wife*, a blog promising to dish on the "dirty truth of life on a gravel road." "I have accepted, enjoyed, and (as of today) love my life as a country mouse," the back-to-the-land blogger opined, "but plunk me down in the heart of Chicago, and I suddenly become *Emily, the city girl*. Give me your traffic! Give me the El! Give me the SHOPPING!!!" In the category of confessions, those surfing the blogosphere in 2011 could indulge in *Secret Confessions of an Amish Farm Wife*, whose About Me page read, "Jacob doesn't know (Lord, forgive me) but I used my egg money to buy this computer. I am somehow able connect through my English neighbor's Internet. Her name is Bunny and she's a bored housewife who drives one of those fancy English cars.

I told her I won't partake of her liquor or tobacco, but when she showed me this thing here called blogging, I was done for. I sneak to the barn and blog during milking hours."

On the radio Rodney Atkins hit number five on the country charts with "Farmer's Daughter." For his music video Atkins, who grew up in Greeneville, Tennessee, played a country boy on TV, getting his real first taste of an "authentic rural life" of "driving a tractor and bucking hay" on the set, according to a news article on theboot.com. Headlined "Rodney Atkins Casts Wife in Hot Video for 'Farmer's Daughter,'" Marianne Horner's write-up described how Atkins got "his real-life wife, Tammy Jo, to star in the video as the farmer's daughter, whom Rodney, a young farmhand, falls in love with and ends up marrying in the storyline." The article failed to mention that in reality, Tammy Jo Atkins made her living at the emphatically less rural occupation of independent dermatological consultant.

Less flattering as an anthem, and evidence of America's increasingly fraught regard for its farm daughters, was Crystal Bowersox's angry 2010 hit for the Sony Records label Jive. The music video, shot in the graying streets of some unnamed metropolis, shows a dingy girl hanging on the sleeve of a world-ravaged mother whose wounded eyes belie existential bewilderment. Anger gravels Bowersox's husky voice as she belts out the real story of a rural mother who breaks her daughter's bones and a dutiful daughter who tells her school counselors that she fell down the stairs. Bowersox choruses with gusto that she's no farmer's daughter anymore. On chat rooms around the country listeners rallied around the agricultural victim narrative, though many assumed the abuse portrayed on the video had occurred at the hands of a farm father. Well into the summer of 2011 the e-debate raged on YouTube, where Sango52 apparently closed the case, commenting, "This is not about her Dad abusing her. It's about her Mom abusing her. That's why they show a woman & a little girl; the knitting represents the unraveling the bad things that happened in the past (i.e. the abuse by her mom), then knitting it into a loving caring life for her son, the abuse stops with Crystal DO U GET IT NOW? (Her mom owns a farm, Crystal grew up on a farm)." Whether its downbeat lyrics referred to a violent ma or pa, Bowersox's song sounded a timeless chord, a blue note likewise sung by writer-musician Debra Marquart in her myth-busting 2006 memoir,

The Horizontal World: Growing Up Wild in the Middle of Nowhere, wherein she writes, "Farmers do not mean to be so possessive; they're just punctuated that way. And farmer's daughters must struggle against the powerful apostrophes of their fathers. They must drive away some early spring morning, hands planted firmly on the wheel, convinced they will never look back."

The forgotten farm child, circa the summer of 2011, somehow crept back into the nation's political discourse, too, serving once more as a poster child in a larger debate. "Why not set as a goal for the 2012 Farm Bill, an ability to add at least 100,000 additional farmers in the area of the small farming and commercial operations?" Secretary of Agriculture and former Iowa Governor Tom Vilsack had earlier asked the US Senate Committee on Agriculture. "Why not establish local advisory councils in communities across the country to identify and recruit and encourage and insist that young people consider a life of farming? . . . And why not expand our efforts to encourage transitions for those seeking to retire and those seeking to start farming? Why not place the nation's attention on our need for young farmers on the same plane as police officers and teachers, as they are equally important for the future of this country."

The problem for farmers' daughters and sons looking to enter the family trade was, in part, credit; without hefty inheritances they couldn't afford to buy the land and equipment, and with lending tightening, the problem seemed to be getting worse. President Obama responded by naming Midwest farmer's daughter Sara Faivre-Davis and two other appointees to the Federal Agricultural Mortgage Corporation, alias "Farmer Mac," in May of 2010, but as currency and market destabilization fears rooted anew and investors turned to solid investments in gold and good midwestern dirt, land prices for would-be Farmer Janes and Joes grew more daunting. In September of 2011 National Public Radio's Clay Masters interviewed sanguine Nebraska farmer Mark Haser, who opined, "If you want to die rich, then become a farmer, because that's about all you're going to do as far as on the rich factor, is you're going to die that way." Sixty thousand fewer farmers under age 35 were reaping and sowing in 2007 than in 1997, Masters pointed out, even though commodity prices were at or near record highs.

The ironies surrounding the nation's farm children and would-be future farmers deepened. Within a few days of Vilsack's call for 100,000 new

pairs of farm boots on the ground, the Tuttles, owners of the oldest contin-
uously operating family farm in America, founded in 1632, called it quits,
citing "exhaustion" of "bodies, minds, hearts, imagination, equipment,
machinery, and finances." When asked by NPR's Michele Norris how the
farm's longtime customers had reacted, eleventh generation owner and
New Hampshire farmer's daughter Lucy Tuttle said simply "tears." The
tears shed by the farm's patrons over the end of an era and the loss of two
iconic farmer's daughters, Lucy and Becky, were born of both happiness
and sadness, Tuttle suggested. In a *New York Times* editorial, farmer and
Times columnist Verlyn Klinkenborg put it more bluntly, challenging the
Tuttle's suggestion that the recession had killed their farm as "too simple."
The economic structure of food and the lack of a level playing field for the
family farm was, Klinkenborg wrote, nearer the truth. "In 1632 and for
many years after, the Tuttle farm was a necessity," the columnist opined.
"In 2010, it is suddenly superfluous, or so we like to pretend."

Two weeks after the USDA released data demonstrating a record number of farmer's markets nationwide—7,175—an increase of over 1,000 from
the year 2010, President Obama and his secretary of agriculture barnstormed
around the upper Midwest in August of 2011, stopping in Peosta, Iowa,
to meet with members of the FFA and other ag stakeholders at the Rural
Economic Forum. While Vilsack spoke to me and other farm writers and
scholars that day of "a record ag year" and "new job opportunities," the
news seemed bleaker for those farm daughters and sons able to access the
Jobs and Economic Security for Rural America report made available to the press
corps. In 2000, it suggested, a daughter or son of an urban American was 10
to 15 percentage points more likely to attend college than his or her rural
counterpart. Meanwhile, the White House Rural Council suggested that the
parents and grandparents of these same rural children now seemed likely
to live shorter lives than their urban cohorts; "mortality rates in urban and
rural areas have diverged," the laconic report read. Not long after Obama
and Vilsack returned to Washington in August of 2011 came news of a US
Department of Labor proposal—the first of its kind in four decades—to limit
child labor on the farm. "Agricultural interest groups don't want federal
regulators to make it harder on them to recruit and train the next generation of farmers," wrote Scott Kilman of *U.S. News and World Report.* At is-

sue were jobs like handling dynamite, chain saws, and pesticides—the kind of hard labor that had earned agriculture the second-highest fatality rate among youth workers, according to the National Farm Medicine Center.

On the other side of the proverbial fence, the many balms of dirt farming achieved traction that bordered on fascination. Gardens, not gadgets, stole headlines in the urbane *New York Times*, including a 2011 story clamoring "Vegetable Gardens Are Booming in Fallow Economy." Americans weren't just returning to backyard "farming" to save money, writer Sabrine Tavernise noted, but to make a point about healthy living and self-reliance to daughters and sons, similar to the lesson First Lady Michelle Obama hoped to instill in digging up the first White House garden plot since Eleanor Roosevelt's. The first lady had never before grown her own, the press reported, but she had made the leap on behalf of her daughters, after environmental journalist Michael Pollan addressed a highly publicized open letter to President Barack Obama as the "Farmer-in-Chief." "My hope," the first lady told the *New York Times*, "is that through children, they will begin to educate their families and that will, in turn, begin to educate our communities." By extension of their mom's convictions, presidential kids Malia and Sasha had become, almost overnight, the Farmer-in-Chief's farmer's daughters.

Elsewhere in popular culture the wider ascendance of young women gave new meaning to the term "cover girl." In a July/August 2010 issue of *The Atlantic* provocatively entitled "The End of Men," journalist Hanna Rosin reported that three quarters of the 8 million jobs lost in the Great Recession were lost by men in industries that were "overwhelmingly male and deeply identified with macho." In education, Rosin pointed out, three young women received bachelor of arts degrees for every two men. Likewise, a September 2010 feature article by *Newsweek's* Andrew Romano and Tony Dokoupil entitled "Men's Lib" documented how the male share of the labor force had declined from 70 percent in 1945 to less than 50 percent, adding that in the country's biggest cities, young, single, childless women earned 8 percent more than their male peers. "It's not how men style themselves that will make them whole again," the authors concluded, citing the decline of the rugged individualism of the farmers, riggers, and welders of generations past, "it's what they do with their days."

But if men themselves were becoming different due the loss of tactile, hands-on occupations like farming, had no one thought to document the corresponding transformation in women? Nowhere in my search could I find an accounting of what might be lost, if anything, in the breaking of a generational chain of leadership-gifted working farm- or ranch-reared daughters that stretched unbroken from First Lady Martha Washington to Supreme Court Justice Sandra Day O'Connor. Thirty-five years after the domestic dust-up and split-ups prompted by US Secretary of Agriculture Earl Butz's ultimatum to the nation's farmers to "get big or get out," and twenty-five years after the farm crisis that landed Middle America in the national news for desperation suicides and bank shootings, what happened to the next-generation country girl whose abiding presence alone might have helped save, or at least mitigate, the loss of the family acres—that girlchild sprung of the working family farm who less than a century ago seemed the epitome of American womanhood?

"It is the country girl that interests us," Martha Foote Crow waxed in 1915, "the promise and hope of her dawn, the delicate, swiftly changing years of her growth, the miracle of her blossoming. There is something about the kaleidoscope of her moods and the inconsistencies of her biography that fascinates us. The moment when she awakes, when the sparkle begins to show in her eyes. . . . We ask no greater happiness than once or twice to catch a glimpse." Crow's starry-eyed description may have been well and good in the era of Babe Ruth and William Jennings Bryan, but what would a catalogue of a twenty-first century farm girl's virtues include? Modesty. Beauty. Self-reliance. Loyalty. Independence. Humor. Good Cheer. Compassion. Energy. Or are traits like these merely latter-day stereotypes of the girl who never was what we imagined her to be, the one that made a habit of defying our cultural expectations? In *The American Country Girl*, Crow seemed to anticipate the question as it might be posed a century hence, and by way of response offered up a "country girl scorecard" based on hundreds of actual letters she received from farm daughters. Under "notable excellences in character," Crow, herself a rural daughter moved from the pampered East to brawling Chicago, listed "integrity, truthfulness, trustworthiness, courage, self-reliance, steadiness, fearlessness, generosity, magnanimity." A partial list of "notable deficiencies" included "inconsiderateness in causing unnecessary trouble to others, carelessness, causing waste or extra work, disorderliness . . . [and] disregarding the rules of home, thus causing worry." In any case, it seemed safe, then as now, to conclude that the farm daughter's was a uniquely fraught inheritance, one strangely out of step with an urbanizing world where traits like "fortitude" and "self-reliance," among the many iconic gifts Crow listed, stand out as virtues sufficiently obsolete or circumstantially altered as to be classed anachronisms. Nor are we culturally naive where the country girl's weaknesses are concerned, weaknesses that prove part and parcel of her rural upbringing. Too often farmer's daughters have been abused, isolated, or deprived; they have sometimes turned hard in the face of the unyielding demands of the exceptional life chosen for, or by, them—become the stereotype embodied in Grant Wood's *American Gothic*.

Who knows how cultural obsession roots, but root it does. And where does it root deeper and more mysteriously than with the farmer's daughter, that chimera that has been captivating Americans since the first yams reared their tuberous heads in Plymouth Plantation. The notion that a profession—be it a colonel, an aviator, a farmer, a UPS man, an astronaut, or Silicon Valley wunderkind—could make our national heart go pitter-patter is not so hard to fathom. But what possible collective heartstrings could be tugged by an appellation earned through a vocational relationship once-removed? No one swoons at the mention of the computer programmer's daughter or the global analyst's son. Tough guy Theodore Roosevelt never said boo of the washerwoman's daughter or the engineer's girlchild, but of the farm wife and daughter, he confidently declared there could be "no more important person, measured in influence upon the life of the nation."

In the era of Roosevelt and the Country Life Movement, the country daughter proved the girl of the hour. For all the rhetoric, it seemed as if the farmer's daughter carried the weight of all Christendom on her shoulders. "God bless the faithful, sacrificing daughters of the farms of our country," wrote C. M. Mather in an October 1898 issue of the *Iowa Homestead* dedicated to the country girl. "Give them every chance to be self-reliant, independent. . . . Remember 'a son is a son till he gets a wife, but a daughter is a daughter all her life,' and she will never disappoint if she has been fairly treated at home." In the same edition Herman F. Jahnke of Regina, Wisconsin, reminded readers, "It is the farmer's daughter that will carry on the world and its work when we are gone beneath the sod and the better we can train her the better she can do." While men focused on the farm daughter's training and her ability to bring forth the next generation, women writing into the *Homestead* sounded a more indignant note. Mrs. Addie Billington of Des Moines wrote, "One almost hesitates to draw the veil and expose to public print a being whose existence provokes so little mention as is bestowed upon the farmer's daughter. It is very common, forsooth, to hear of the wonderful career of some statesman, official, magnate, or millionaire who was a 'farmer's son,' but not often does the comment revert to the farmer's daughter." Mrs. Billington was right; famous men had penned an entire genre of books for sons of the farm, as exemplified by publishing icon Henry Wallace in his *Letters to the Farm Boy*, which, spurred by reader demand, went through five editions beginning with its first printing in 1897. "Out from among boys like these have come renowned authors, statesmen, and presidents of our vast country," read the annual report of the New York State Agricultural Society in 1891. "But what of his sister, the farmer's daughter; has she no birthright, no part in the grand march forward that the world is surely making?" Calling "the rural question" the "central question of the world," Crow agreed, intending no hyperbole in venturing, "If the social problem is the heart of the rural problem, and if the failure of the daughter's joy and usefulness threatens the farmstead— then once more in the history of the world has the hour struck for woman; then does the welfare of the world depend upon her as much as did the life of the bleak New England shore depended on the health and survival of the Pilgrim Mothers."

TWO YEARS BEFORE I was born in 1973, the Senate narrowly approved President Nixon's appointment of Earl Butz to be secretary of agriculture, and midwestern farmers braced themselves for a bloodletting. On hearing the news, the Iowa National Farmers Organization (INFO), the November 22, 1973 edition of the *Caroll (IA) Daily Times Herald* reported, introduced a resolution charging the secretary of agriculture designate with being an "enemy of the small farm." Butz, who had since the 1950s preached the "get big or get out" and "adapt or die" rhetoric of his predecessors, was viewed as an advocate of corporate farming. His appointment, farmers feared, would effectively kill the *family* in *family farm*, striking a death blow to the next-generation farmers' daughters and sons hoping to take the wheel of the John Deere or International Harvester. Referring to Butz's "adapt or die" and "get big or get out" mantras, farmer-writer Wendell Berry later wrote in the *New York Times*, "Those are two of the most ignorant, undemocratic, and heartless things ever said to the free citizens of a democracy."

On our midwestern farm, however, prospects seemed brighter, the dream still more or less intact. Two of the family's three farm daughters lived within a quarter mile of the home place. My aunt Barb had set a wedding date for the summer of 1975. My mother, herself a farmer's daughter, had recently returned with my father to his parents' farm. For a time all of us—grandparents, parents, aunts, and a passel of cousins—shared common ground. My grandmother Julia, a farmer's daughter turned farmer's wife, served as our ringleader and queen entrepreneur. In the previous decade she'd begun her own ceramics studio, Coral Gables, in the basement of her farm home, from whence she taught pottery classes to the women of our rural community. In 1972 she'd mustered the country women of my family together in a joint venture that made the front page of the *Cedar Rapids Gazette*, where a photo spread ran bearing the caption: "Everybody works who visits the site of Wind in the Walnuts near Mechanicsville. Busy scrubbing the loft of the big red barn are, from left, Mindy Coon, 10, Tasha Jack, age 15 months, and Rodney Sullivan, 6, three of the six grandchildren of the Ed Jacks."

In those days it was customary to address the farmer's wife as "Mrs.," as in "Mrs. Ed Jack," but the truth was my grandmother, rather than my grandfather, was the driving force behind the headline "Art Show on Farm Is Iowa Family Project." In a second photo my grandmother smiled demurely in front of an antique stove she'd just wiped clean, the caption billing her as "director" of our enterprise. Datelining his feature "Mechanicsville, Iowa," newsman Art Hough began, "It's only a few miles south of bustling Highway 30 to the site of Wind in the Walnuts, but it might as well be a hundred. Wind in the Walnuts will be a country art market . . . on a farm whose

buildings go back more than 100 years. . . . A family project, directed by
Mrs. Ed Jack, the diversified art show will be produced under the shade of
a grove of beautiful walnut trees and in a big red farm building which has
been reconverted from a woodworking shop to what will be a year-round
art center." Ours was what even then was considered a "feel-good" story
about a family who somehow managed to hold it together. "The reason
Mrs. Jack calls this a family project is because it is," Hough insisted. "The
Jack's three daughters, a daughter-in-law, and even the six grandchildren
are hammering away and sawing, scrubbing, painting, and polishing to get
the barn and grounds ready for the public."

What made Hough's story front-page material was the real headliners
in my family: its farmers' daughters. "All of the family, at least the women,
are artists in their own right," the article trumpeted. Going on to list my
grandmother's cadre of she-assistants, the article read, "Assisting Mrs. Jack
in the art show preparation are the Jack daughters, Mrs. Charles Coon,
Mrs. Robert Sullivan, and Miss Barbara, and a daughter-in-law, Mrs. Michael Jack. All but the Sullivans, who reside at 706 Danbury Street SE,
Cedar Rapids, live in the immediate area." The object of the Wind in the
Walnuts, according to my grandmother, was "to get people in this area interested in their own local artists."

The human interest story of our farm daughter-led venture was likewise picked up by the Iowa City Press-Citizen, where word of the "first annual
country market" seven miles southeast of Lisbon appeared beside movie
ads depicting the broader gender fires then reaching combustion in the
world beyond the barnyard. That same weekend in 1972 the Englert Theater in Iowa City featured Jack Lemmon and Barbara Harris in The War
Between Men and Women, while at the Cinema II at the Sycamore Mall appeared a short-skirted, knee-booted Jacqueline Bisset in Stand Up and Be
Counted, a flick whose tagline read, "From Adam's Rib to Women's Lib
. . . baby, we've come a long way." Over at the venerable Iowa Theater The
Trojan Women was being shipped out to make room for Dustin Hoffman's
The Graduate. And, finally, in the pipeline at the nearby Stage 4 waited a
movie that would spook a generation from a life in the boonies, John Bor-
man's Deliverance, a soon-to-be classic that would lend new and trenchant
meaning to the phrase "I bet you can squeal like a pig."

For her part my grandmother was too busy with the work of organizing her agrarian arts fair to lose much time contemplating Hollywood versions of depraved hillbillies or aggrieved women's libbers. The same held for my mother and my aunts, whom area newspapers pictured swinging hammers and wielding paint brushes instead of burning bras or holding picket signs. Our women-empowered operation seemed bigger and better than ever, having mushroomed from 100 acres in 1917 to nearly 500 by 1974, when my father left his job as the manager of the Iowa Falls, Iowa, municipal airport to help my grandfather farm. My father had voted with his feet at precisely the hour of greatest need for a midsized family farm like ours in a profession the Associated Press called in 1971 "one of the biggest gambles on earth." Even then, farming was, by the numbers, a losing proposition, with extension agents all over the Corn Belt noting that it cost $1 to produce a bushel of corn in years when farmers were getting less than a dollar a bushel for their troubles.

None of us yet knew in 1973 that the decades-long run of the farmer's daughter and the farm she helped sustain would soon come to an end. Three months before my mom and dad swaddled me up and carried me out into the cold November world, Denise Rupp, Iowa's favorite Farmer's Daughter of 1973, had appeared just up the road from us at Armstrong's department store, a fixture in the old Czech town of Cedar Rapids. "Come in and meet lovely Denise Rupp in the downstairs store," read the ad in the hometown *Gazette*. "Denise is appearing in conjunction with the Downtown Agriculture Appreciation Days. Visit her between 10 a.m. and 4:30 p.m." At that time the Midwest still broke for its farmers' daughters. Even the most stoical bachelor couldn't resist the image of fair Denise, dusky blond hair falling at her shoulders, tiara perched winsomely atop her head, smiling at us wholesomely from the newsprint while, just south of her lovely visage, Famous Name bras were going for $1.79 and Holiday Girdles could be had for $4.99.

The pageantry of the historical moment belied a greater disconnect between the fading yet still beatific image of the farmer's daughter and her on-the-farm reality, for even as Butz's approval ratings dropped to 23 percent among Iowa farm people in a *Wallace's Farmer* survey, country life seemed to be clipping along quite nicely in the mass media. On television we farm kids thrilled to the countrified exploits of *The Dukes of Hazzard* as they fought the townie powers-that-be each Friday night as we munched on Party Pizzas. On Saturdays *Hee Haw* and *The Lawrence Welk Show* reached us blissed-out country bumpkins in what seemed like perpetual syndication. On shows like these being rural was still fun and funny. Farmer's sons were still earnest and punny and farmer's daughters were beautiful, if not crafty. Little did we know that urban America had already moved from rural stereotypes to rural indifference, so much so that in 1971, the year my sister was born into farm daughterhood, CBS nixed from its regular network lineup *Hee Haw* and its country cousins *The Beverly Hillbillies*, *Mayberry R. F. D.*, and *Green Acres*—citing the network's feelings that the shows' viewers reflected the "wrong" demographics (i.e., rural, older, and less affluent), even though the shows remained popular in the ratings. Still the "Rural Purge," as it would come to be called, hadn't seemed to reach our local Cedar Rapids affiliates. Thanks to the wonders of syndication, Lawrence Welk didn't have gray hair, and Buck Owens and Roy Clark, we imagined, were still as popular on State Street and Wall Street as they were at Sutliff Bar and Grill. On the radio and jukebox our misconception that our agrarian life was normal was reinforced by the bevy of 1970s farmers' daughters that Nashville feted in its country music award shows: Loretta Lynn, the bona fide farmer's daughter from Butcher Holler, and the one-room-cabin-reared Dolly Parton, who, as it happens, was the number-one, heart-pounding crush among the male country dwellers of my boyhood, and the subject of several risqué jokes told by my oldest, boldest cousins.

While nary a platinum blond was to be found on either side of my family tree, it turns out we had more in common with Parton than *The Jeffersons* or the *Mary Tyler Moore Show* or any of the more urbane sitcoms CBS drafted to replace rural television. No matter how done-up or surgically enhanced, Dolly was one of us. In interviews she recalled growing up in a one-room cabin on the banks of the Little Pigeon River. She spoke of deep poverty and compensatory creativity, remembering how she and her five siblings made homespun fake tattoos out of the purple juice of pokeberries. Too poor to afford makeup, Dolly used the tips of burned kitchen matches to make a paste for eyeliner, pokeberries for rouge, and flour for base, a concoction about which her sharp-tongued mama once joked, "What are you gonna do when you get hot and sweaty, break out in biscuits?"

"Womanhood was a difficult thing to get a grip on in those hills unless you were a man," Parton later wrote of her farm daughter days on a small plot of corn ground outside Sevierville. "My sisters and I used to cling desperately to anything halfway feminine. For a long time I was a tomboy, but once I got a better idea of what it meant to be a woman, I wanted it with everything in me." Likewise, in our midwestern home our thrifty mother, like Parton's, directed my sister and me to use newspaper in lieu of that ever-scarce commodity, toilet paper. Parton, meanwhile, put the daily news to use as wallpaper to keep the drafts at bay. "We could see the pictures of the models in the newspapers that lined the walls of our house," Parton recalled in her memoir *Dolly: My Life and Other Unfinished Business*, "and the occasional glimpse we would get at a magazine. . . . They didn't look at all like they worked in the fields. They didn't look like they had to take a spit bath in a dishpan."

Like Parton's, our family in the early years worked together in the fields to make things go. Though the bulk of the work fell to my father and grandfather, my adult aunts, the nine grandchildren, and a couple of in-laws walked beans and gleaned corn in season and regularly helped pick rocks from the ditches, fields, and watercourses. We made farm art together, too, learning from our grandmother that creativity meant reuse—Uncle Lee's old horse and hay barns recast as a rural arts center; my grandpa's cans of Yuban coffee repurposed for everything from containers for nuts and bolts to a repository for the family's loose change and extra buttons. For extra income my artistically gifted mother stitched the sock monkeys that would become iconic of the 1970s, fashioned delicate dolls from corn husks, and made wreathes from the roadside wildflowers and herbs that we kids harvested at her bidding and sold alongside her at local farmer's markets. Though our Grade A Iowa loam made our existence far less hardscrabble than Parton's, we understood what Dolly meant when she wrote, "The whole family might work for days to clear trees or move rocks or whatever it took."

Perhaps because of the blinders we'd strapped on to make our rural lives work, we could not see the larger societal forces changing farming forever, changes that were making my sister and me the last of the family's many generations to be raised a farmer's daughter and son, respectively. When cars unmarred by gravel and lime sometimes slowed on our country road to watch us at work or at play, we never stopped to imagine that we might end up in some documentary photographer's funereal viewfinder, or some budding scholar's college-ruled field notes. Little did we know it, but we roughhousing, rumpusing 1980s farm children had all but entered the history books as fodder for woebegone supposition by scholars. "The farm today is mainly a place where people grow up," opined the foremost rural scholar of his generation, David Danbom, in *Born in the Country*. "When they have reached adulthood, they are more likely to leave and become accountants or computer programmers or retail clerks than they are to follow in their parents' footsteps."

Even the analysis of the problem seemed to reveal a generational if not gendered gap. In the era of Earl Butz and later, male scholars of agriculture tended to focus on the mathematics of the crisis—the economic bottom line—as the reason for the family farm's demise, and on the male operator as the rainmaker in an ungodly economic drought. Two generations earlier scholars including noted farmer's daughter and university professor Mary Meek Atkeson focused instead on the matriarch as the prime mover of an agrarian future. "If she [the farm woman] succeeds, she will have the satisfaction of having her children love the farm and the farm life," Atkeson wrote in *The Woman on the Farm*, "and the country dwellers of the future will be descended from the same hardy and vigorous stock. . . . But if she fails, then our American country life as we have known it and loved it—the beautiful and inspiring and hearty country life of Washington and Western pioneers—will be a thing of the past."

Everywhere our world was changing, though change proved paradoxical. On one hand, all was as it had been since 1855—the family ground securely in our hands, corn yields rising, another only farmer's son—me—ready to inherit his father's trade in a decade or two. In 1976, when I was 3 years old, the Iowa Department of Agriculture issued us a certificate that seemed to all but seal the deal, recognizing ours as a "Century Farm." "This family, having owned, and been the stewards of Iowa land for over 100 years, has significantly contributed to the growth and stability of Iowa agriculture, truly making Iowa 'A Place to Grow,'" it read.

It was America's bicentennial year, and a proud family snapped photos of us kids playfully dressed as Yankees and Redcoats for parade floats made to mark the occasion. Nixon had been ousted, American boys had returned from Vietnam, and a peanut farmer, Jimmy Carter, had been elected to the nation's highest office. Corn prices were high in 1974 and 1975, close to four dollars a bushel, a rate that, adjusted for inflation, would net today's plowman $14 or $15. Butz's USDA had made a measure of prosperity in the corn market with its "fencerow to fencerow" policy, as average farm families like ours felt as if we were expanding, even if we hadn't purchased a single extra acre. In short, we, along with many Heartland corn-growers, were riding high on the hog and feeling pretty patriotic.

We were big, and we were beautiful. When I was 6 months old in July of 1974, one of our own, Iowa farmer's daughter Rebecca Ann King, achieved the nation's highest office of beauty, earning the Miss America crown and setting off to tour Middle American department stores, business conventions, and state fairs. King's appearance at Choate's department store in Winona, Minnesota, occurred at a curious cultural moment, a space and time when the women's rights movement and the impending decline of the family farm seemed to join in the frame of a 5 foot 9-inch, 125-pound, blond farmer's daughter, who aspired to become a juvenile court judge after completing her reign and collecting on more than $100,000 in appearance money. Post-Watergate, a farmer's daughter holding down the nation's highest, hottest office attracted the ironic interests of even the *New York Times* NewsService's Judy Klemesrud, whose article "Miss America: A Busy Iowa Girl" was picked up by the *Ames Daily Tribune* in July of 1974. Klemesrud quoted with apparent relish 16-year-old Miss America autograph-seeker Nancie Pickett, who sour-graped, "I don't think she's so great. I don't think Miss America should say that she won't sign autographs until people get in a line. She acts like she doesn't have time for us." Likewise, Robin Moe, age seventeen, wondered aloud, "I don't see why she has to wear false eyelashes and fingernail polish and be so phony." Klemesrud herself managed to corner King over a tuna sandwich in the backroom of Choate's long enough to ask the reigning queen what she thought of two pressing issues of the day: the women's movement and Watergate. Klemesrud quoted Miss America's seemingly naive answers with barely contained glee. "I really feel Watergate has been good for us," King told the *Tribune*, adding for good measure this political hot potato: "If a woman wants to stay at home, fine, and if a woman wants to be a lawyer, fine. Maybe I'm not as radical as some."

Middle America, it seemed, had woken up from its sexual torpor, Rip Van Winkle-like, and wanted not only to throw the rascals out in Washington, but to chuck in the dustbin the other shopworn, patriarchal institutions that went with them, farmers' daughters included. Almost overnight America's real-life country girls found themselves under fire on all fronts—economic, cultural, spiritual, sexual. Wasn't the term itself—*farmer's daughter*—evidence of objectification, feminists asked. What happened when a so-called farmer's daughter grew up and became something else? Could one ever graduate from being a farmer's daughter? Wasn't the farm daughter, if in fact she lived and worked on the farm, another of the ravaged victims of patriarchal oppression and environmental degradation, the equivalent of unpaid slave labor toiling away in a business in which she would never be a stakeholder, let alone a shareholder?

During the very year that America chose as its defining miss a Midwest farmer's daughter, the man who would become the nation's foremost contemporary agrarian, a Kentucky horse farmer named Wendell Berry, was formulating his definitive book, *The Unsettling of America*, a treatise in which he challenged the reactionary feminism that would, if it could, jettison both Miss America and traditional homemaking. In his essay "The Body and the Earth," Berry examined the phenomenon of sexual division through the commonsensical lens of the yeoman. Critiquing the modern, off-the-farm marriage as "two careerists sharing a bed," Berry wrote, "the modern failure of marriage that has so estranged the sexes from each other seems analogous to the 'social mobility' that has estranged us from the land."

As I grew into adulthood national and local coverage of the farmer's daughter shifted from the propagandistic boosterism of the 1950s to anything-you-can-do-I-can-do-better triumphalism of the 1970s. A case in point was an AP article datelined Fonda, Iowa, that told the story of Rhonda Maloureux, who had, as part of the back-to-the-land movement then underway, dropped out of stewardess training in Kansas City for a job, as the article put it, "working the good earth." "I didn't like being cooped up all day," Maloureux said of her reasons for leaving a high-flying career for a decidedly unglamorous job at George Fulcher's beef farm. "Some things get kind of heavy for me," she confessed, adding, "but I have done all kinds of fieldwork except planting." Maloureux wouldn't disclose her salary, but she made, the article felt obliged to point out, "as much as any man holding a job like this." Duly impressed, writer Emmett Butler declared in his column in the *Treynor (IA) Record*, "Agriculture has made much progress in many areas in the past generation, with one advance which is particularly noteworthy. Nowadays you can't tell a farmer's daughter from a city gal." At the time no one stopped to ask what would be the cultural loss when the farmer's daughter traded in her crown of thorns for a reign of invisibility.

CHAPTER TWO

THE MIDWEST FARMER'S DAUGHTER

IN THE SUMMER of 1965 the Midwest farmer's daughter was still a song worth singing. On a momentous LSD trip the Beach Boys' Brian Wilson and Mike Love had written a ditty that would go on to be ranked by *Rolling Stone* in the top 100 greatest songs of all time, and by the Rock and Roll Hall of Fame as among the top 500 to shape rock and roll. The song was "California Girls," an ode not just to the girls of the Golden State, but to the Midwest farmer's daughters, who, the boys crooned, could make you feel alright.

In August of that same year Chubby Checker instructed my 15-year-old mother and tens of thousands of others farm girls on the intricacies of the "The Twist," "The Mess Around," and "The Fly." Among the other dances sweeping the nation that summer were those with names a country girl might understand—"The Pony," "The Mashed Potato," and "The Chicken," though perhaps "The Blue-beat Fly-away" best expressed her growing desire to join the bobby soxers in town. By the time the Beatles arrived in 1964, and found my Aunt Barb readying herself for graduation from Mechanicsville Consolidated High School, the notion that the rock and roll era had reached the region's farm sons and daughters became apparent at a glance. "All the boys were wearing bell bottoms and letting their hair grow long," Barb recalls. "It was cool, man!"

IN 1965, HAD you been searching for a country musician, you'd have invariably thumbed your way to Nashville. If you'd been canvassing for a coal miner's daughter, you'd have plotted a course straight for the hills and hollers of West Virginia. And if you'd been trekking for a tried-and-true farmer's daughter, you'd have doubtless mounted your Pinto and pointed it toward the heart of the Heartland. Judges and jurors of such things anointed Iowa queen commonwealth of the country gal dating back to the first of many national farm daughter contests staged during the baby boom. In 1948 the United Press newswire hummed with the bulletin that Pasty Miller, age 17, of Osceola, Iowa, had been crowned the nation's prettiest farmer's daughter at the National Farm and Garden Show in Chicago. Running close behind was a rural route head-turner from north central Indiana, and, further down the list, was Osceola, Iowa's, Dorothy Lacey, who had earlier knocked off her fellow hometown rival for the title of Iowa Soil and Conservation Queen before drawing short straw to her soil sister at nationals.

"Thousands of city folk crowded into the huge livestock show to watch the best products of the nation's fields judged," the United Press story crowed, making explicit what many in farm country had long thought: a crop is a crop is a crop, whether a corn-raised girl, boy, or livestock. No ironic titters mocked the holders of such titles, nor did Corn Belt daughters turn russet in shame at being named "Pork Queen." Being judged the country's fairest farmer's daughter was an undiminished, if not unambiguous, honor, akin to the pride George Saureman might have felt when he claimed the prize that same year for best hay shown. Back then pageants widely captured the imagination of both rural and urban America in what amounted to an unusual, and ultimately fragile, symbiosis: Middle America produced and displayed its best "crops," and urban American turned out to gawk.

A mere year before Patsy Miller donned the sash and tiara in Chicago in 1948, my grandfather, Edward Lee Jack, had taken second in the national plowing contest as crowds of 40,000 plus—as many as today might attend a home game in the heart of Big Ten football country—looked on, rapt. When the dust settled, Edward Jack was awarded his second place trophy by that year's Plowing Queen, Carolyn Wiese of nearby Bennett, Iowa, who had been presented her crown by Iowa's junior US Senator Bourke B. Hickenlooper, thus begetting the unforgettable headline "Hick Crowns Queen." It had been a clean sweep for the Hawkeye State—first and second in clean plowing and first in beauty thanks to the fresh-faced Weise.

If popular farm contests served as any indicator, Iowa, circa my grandfather's silver-medal performance behind the plow, had reached the absolute apogee to which any agrarian civilization could dare hope to aspire, balancing artistic and agricultural excellence in a rare, one-two punch. Years prior, Iowa City, Iowa, had been declared the "new US literary center" by Boston literary critic E. J. O'Brien. The state as a whole had bested all comers in production of corn and hogs and had produced a bevy of clean plowing and corn picking national champs, including my grandfather. Its native son and patriot, Glenn Miller, had gone platinum with his album *Chattanooga Choo Choo*, and in 1947, the year Ed Jack hoisted his trophy, another native son, Pioneer Hybrid founder Henry A. Wallace, had announced a run for president on the Progressive ticket. Wallace promised not just farmers, but the whole of America, an abiding agrarian belief in the "old-fashioned American doctrine of standing up, speaking your mind, and letting the chips fall where they may."

Meanwhile, J. Edward Kirbye, vice president of the Iowa Press and Authors' Club and later president of Drury University, had become sufficiently besotted with this remarkable commonwealth and its achievements to publish his "Creed of Iowa," a state-specific pledge of allegiance that, like the annual farm contests, yoked the state's superior farm products with its Grade A crop of human resources—its farm girls and boys. "I believe in Iowa," read the pledge, "land of limitless prairies with rolling hills and fertile valleys, with winding and widening streams, with bounteous crops and fruit-laden trees, yielding to man their wealth and health. . . . I believe in Iowa, rich in her men and women of power and might. I believe in her authors and educators, her statesmen and ministers. . . . I believe in Iowa, magnet and meeting place of all nations fused into noble unity, Americans all, blended into a free people. I believe in her stalwart sons, and her winsome women . . ."

It was the state's finest products, especially its winsome women, that Cedar Rapids's legendary farmer radio station, WMT, sought to highlight when it cosponsored the National Clean Plowing Contest and, a decade later, the Iowa's Favorite Farmer's Daughter Contest. From the late 1950s through the late 1970s, when declining applicants and changing cultural mores effectively ended the annual tradition, Iowa's farm families yearly read Cinderella calls like this one from the 1963 *Ames Tribune:* "Somewhere in Iowa is a comely farmer's daughter who through her appearance, personality, and background may win a trip to Washington DC as well as prizes. Girls are reminded less than two weeks remain for entries in the sixth annual search for 'Iowa's Favorite Farmer's Daughter.'" The invitation further offered, "Any Iowa girl who is single, between 16 and 21, and who has a definite farm background is eligible."

The contest clipped along quite nicely in its first decade, routinely turning up Heartland girls who had grown up choring on working grain or livestock farms and who had been stalwart members of their local 4-H and FFA chapters. In 1967 Carol Wilson of West Liberty, Iowa, won the trip for two to the nation's capital, a $300 dollar wardrobe, a portable color TV set, a transistor radio, and a set of encyclopedias. It was an enlightened gift package—one that, beyond the wardrobe, would no doubt have pleased a favorite farmer's son had there been a contest staged in his honor. The *Burlington Hawk-eye* shared the winner's measurements without a whiff of awkwardness. "Miss Wilson is 5 feet 8 inches tall, has blonde hair and blue eyes," the article crowed, further describing her as a gee-whiz choir member of an a capella group and a spirited lead in an unnamed folk group. In the summer of 1967 Wilson appeared at the Old Thresher's Reunion in Mount Pleasant, Iowa, and later that weekend loosened up her angelic vocal chords as a special guest on the Giant Cavalcade of Power on radio station KXGI.

That same year Middle American audiences tuned into *The Farmer's Daughter,* a half-hour comedy series about a rural miss from Minnesota named Katy Holstrum who, while working as a governess for a congressman in Washington, DC, falls head over heels for her legislator boss, Glen Morley, played by William Windom. In the premier episode on ABC, entitled "Here Comes the Bride's Father," Katy and Glen, returning from the White House, "are surprised by her parents who have arrived from Minnesota to discuss wedding plans." All's well until the spoilsport Papa declares that the wedding must be held back in the Lutheran climes of "Yasperville," Minnesota. In 1967 such come-home bubble bursting qualified as tragicomic plot twist.

By 1971 the Iowa farmers' daughters who had once ruled the roost of local and state pageants had become rare enough to merit front-page coverage when, per chance, they achieved a breakthrough victory. When Judy Faye Stephens was crowned Miss Iowa on June 18, her hometown of Clarinda was justly proud, posting signs proclaiming "Farmer's Daughter Makes Good." A picture showing the 21-year-old Stephens smiling demurely and holding two precious kittens appeared on the front page of the *Des Moines Sunday Register* along with her measurements ("36-23-35"), her hair color ("honey blonde"), and her skin type ("peaches and cream"). For *Register* staff writer Nick Lamberta, the victory "testifie[d] to the advantages of outdoor living on a Page County farm." Stephens, one of five children of Mr. and Mrs. J. W. Stephens, neither smoked nor drank, and she spent her Sundays away from the family's 700-acre farm teaching Sunday school at the Church of Christ in Hepburn, Iowa. Her mother would substitute teach for her daughter while she took time off to attend the Miss America Pageant in Atlantic City, the article intimated. And if all that didn't justify a dozen column inches on the front page of the state's leading newspaper, the pride of Clarinda "was one of the few contestants who didn't wear false eyelashes." Lamberta remarked, "Miss Iowa told the press, 'My own eyelashes are thick enough, and if I try to put false ones on they got all tangled up,'" adding that while she was no tomboy, she didn't mind helping her dad with chores or her mother prepare meals. She was no rube, either, having declared a history major at Morningside College in Sioux City and announced her intentions to pursue a master's degree. When Stephens's hometown fans turned out for a reception in her honor that summer of 1971, they came bearing placards reading "5 foot 7, Judy's Heaven." The sloganeering seemed apt.

While a generation earlier the superiority of farmer's daughters had been taken for granted by many in farm country, now a God-fearing farmer's daughter winning a gold medal seemed just shy of miraculous.

IN RETROSPECT THE songs atop the Billboard charts that sultry summer of 1965 weren't songs meant for dancing, but ballads or anthems penned to remind the last vestiges of 1950s America what was right with the republic before it changed forever.

In 1966 *Cedar Rapids Gazette* farm writer Rex Conn traveled out to Mechanicsville, Iowa, to talk with my great-uncle, George Brown, about the struggles inherent in keeping up a 486-acre, 70-head-of-cattle family farm with his own children gone. In many ways George and Mary Brown, née Mary Puffer, had an enviable problem—their talented, intellectually gifted children had found employment and educational opportunity awaiting them across the country. Of the seven Brown children, one had landed a job at a regionally famous radio station in Ames, Iowa; another had earned a PhD in chemistry and begun his own successful business in Fort Lauderdale, Florida; and still another had been recruited to St. Louis to work for Douglas Aircraft as an electronics technician, leaving George to pick most of his 110 acres of corn by himself. "It's practically impossible to get help," Brown lamented to the big-city reporter. Indeed, that same week in 1966 the *Gazette* reported that the number of hired farmworkers in 1965 had declined 7.2 percent according to the USDA.

While the farm rolls plummeted, sales of several albums hit new heights in the summer of 1965. "It was a mad, topsy-turvy scramble in the singles race," Jeanne Harrison observed in the Elyria, Ohio, *Chronicle-Telegram* "Platter Patter." The Beach Boys, Harrison reported, had "come out of the pack" to register seventh in the national hit parade with "California Girls," beating out the Beatles's "Help" and Billy Joe Royal's "Down in the Boondocks."

The Beach Boys struck many as odd messengers for the wishes and dreams of Middle America's country Jacks and Jills, but the boys, like Lawrence Welk before them, had capitalized on a romantic geographic heritage in order to tap a national vein. In fact, as Brian Wilson told AP movie and

TV writer Bob Thomas one year later in a 1966 piece appearing in the *Mt. Vernon (IL) Register-News*, "Our dad was a songwriter who had a number recorded by Lawrence Welk and did alright with another song." The song penned by Murry Wilson was called the "Two-Step Side-Step," and Welk and his orchestra had played it live on weekly radio in 1952. Ironically, the the "Two-Step" had much more of a country-western, Middle American feel than the fun, floating harmonies of "California Girls." More ironic still was this fact: while only one of Murry's boys surfed, the boy band had somehow become the international symbol of surf culture. That a group of blond-haired, blue-eyed Californians would lend the term "Midwest farmer's daughter" national currency seemed consistent with the paradox that found a song about a regional cipher climbing the charts at exactly the moment when in real life she found herself packing her bags for a place like California. By 1970, reported historian Jon Gjerde in his essay "Middleness and the Middle West," nearly 40 percent of native Iowans had left their home state. A 1960 survey deepened the irony, showing most of the newcomers to Los Angeles had been born in Minnesota, Iowa, Missouri, the Dakotas, Nebraska, and Kansas.

BRIAN WILSON WAS far from the first male writer to capitalize on the romance of a farm daughter fly-by. For more than a hundred years the hard-working Middle American girl had been penned as a muse or ingénue to be appreciated symbolically, and at safe remove. One of the earliest front-page drive-bys appeared in the *Rural Repository* of June of 1844, accompanied by an illustration of a farm girl in bonnet and gloves smiling coquettishly from horseback. Writer William Howitt opened his romantic travelogue with a warning to the unsuspecting gentleman:

> As you are thinking only of sheep or of curds, you may be suddenly
> shot through by a pair of bright eyes, and melted away in a bewitch-
> ing smile that you never dream of till mischief was done. In towns,
> and theaters, and thronged assemblies of the rich and the titled fair,
> you are on your guard; you know what you are exposed to, and put
> on your breast-plates and pass through the most deadly onslaught of

beauty—safe and sound. But in those sylvan retreats, dreaming of night-ingales, and hearing only the lowing of oxen, you are taken by surprise. Out steps a fair creature, crosses a glade, leaps a stile; you start, you stand—lost in wonder and astonished admiration; you take out your tablet to write a sonnet on the return of the nymphs and dyrades to the heart, when up comes John Tompkins and says, 'It's only the Farmer's Daughter! . . . I tell you they have such daughters. Those farmhouses are dangerous places. Let no man with a poetical imagination, which is but another name for a very tenderly heart, flatter himself with fan-cies of the calm delights of the country; with the serene idea of sitting with the farmer in his old-fashioned chimney corner, and hearing him talk of corn and mutton—of joining him in the pensive pleasures of the pipe, and brown jug of October; of listening to the gossip of the comfortable farmer's wife; of the parson and his family, of his sermons and his tenth pig—over a fragrant cup of young hyson, or lapt in the delicious luxuries of custards and whipped creams. In walks a fairy vi-sion of wondrous witchery, and with a curtsey and a smile, of most winning and mysterious magic, takes her seat just opposite. It is the Farmer's Daughter! A lively creature of eighteen. Fair as a lily, fresh as a May dew, rosy as the rose itself; graceful as the peacock perched on the pales there by the window; sweet as a posy of violets; modest as early morning, and amiable as your imagination.

Howitt, in fact, sings something akin to his very own, indefatigably sexist "California Girls," offering a taxonomy of the farmers' daughters who might beguile the heart, just as the Beach Boys split hairs concerning "the cutest girls in the world." Howitt's taxonomy perpetuated several enduring ste-reotypes of the farmer's daughter. The first class, the columnist stipulated, were the daughters of what he called "men of estates and large capitals," the landed elite and hobby farmers who had made their money in a far-off city while hanging their hat in a country manse. The daughters of these men, observed the antebellum scribbler, were really not that much different from other girls in their dislike of hard work. These products of boarding schools and elite colleges, these drawing-room flowers, the author wrote, could not, despite qualifying in the broadest sense for the appellation, be persuaded to "leave the piano for the spinning wheel." Less desirable for the young squire or courtier was the country girl Howitt allegorically dubs

"Dolly Cowcabbage." To the chagrin of her neighborhood's many bachelor farmers, Dolly was "a little stout plodding woman with a small round rosy face." As if in afterthought, Howitt added, "She has offers: men know what's what, though it be in a homely guise; but she only gives a quiet smile, and always says 'No! I shall never marry while father lives.'"

The final type in the lineup was the tomboy, the woman who "as a girl romped and climbed and played with the lads of the village. She swung on gates and rode on donkeys. . . . She thrashed the dogs, fetched the eggs, saw to the calves, and then mounted on the wall of the garden, with her long chestnut hair hanging wild on her shoulders, and a raw carrot in her hand, which she was ready either to devour or to throw at any urchin that came in sight." A wild and fearless creature in the author's eyes, such a girl would inevitably be sent away to live with an aunt in the city, who would see to it she would have a proper "bringin' up." Readers would know this specimen at a glance, reminded the newspaperman, on account of a visage in which lived "a mixture of life, archness, freedom, and fun that . . . was especially attractive and dangerous to look upon." Her eyes especially, Howitt noted, belied a chimera's nature, orbs of a "half-a-dozen different colors, if half-a-dozen people different people might be believed." Everywhere this born-on-the-farm, refined-by-the-city farmer's daughter went, the eyes of admirers inevitably followed. "The gay young gentleman farmer, the rich miller, the smart grazier, the popular lawyer of the country town, were all ready to fight for her," the scribbler insisted, reminding his readers that she invariably had her sight set on a "younger and handsomer husband," an upwardly mobile colonel destined to take her away.

Historically, interest in the farm daughter spiked in the decades following the Luddite rebellions of the early 1800s and the Swing Riots of 1830s Britain, in which farmers destroyed threshing machines in protest of widespread mechanization. By 1850 the company town of Lowell, Massachusetts, begun as a utopian experiment for worker rights, had, ironically, turned into the largest industrial complex in the nation. And as the smokestacks and smelters of factory towns like Bethlehem, Pennsylvania, reached ever closer to America's farmland, the dream of Arcadia and its daughters captured the Western world's romantic imagination several centuries after Sir Philip Sidney's long poem of 1590, *The Countess of Pembroke's Arcadia*, resurrected Virgil's Arcadia from *Eclogues* and lent it new life as a romantic trope. Connecting the natural beauty and fertility of the land to the buxom beauty of its women, Sidney waxed, "It must needs be, that some goddess this desert belongs unto, who is the soul of this soil, for neither is any less than a goddess worthy to be shrined in such a heap of pleasures, nor any less than a goddess could have made it so perfect a model of the heavenly dwellings."

As industrialization tightened its grip, the farmer's daughter found herself a frequent conceit in antebellum American periodicals, where treatments harkened back to classical myth. America's disappearing farm daughters, sisters in a circumstantial sorority, presented "dangers of no trivial description," Howitt claimed, "that haunt the bush, though there be no lions . . . the Farmer's Daughter in the bloom of beauty is not to be carelessly approached. She can sing like a Syren, and is as dangerous as Circe in her enchanted island." Howitt and others read into the farmer's daughter a particular danger for a generation of urbanizing American men who might be tempted by such sylvan visions to swap their briefcases back for saddlebags, thereby returning the de facto shares they had bought in a nascent and still vulnerable industrial system.

"There is no place in the world more joyful and quiet than a farmer's home," D. W. Bartlett begins his parable in the 1851 *Ladies Wreath* annual. Here, the author insisted, one could hear the "sweet voice of the farmer's daughter singing happy songs, artless but beautiful," and there assuredly would be "springs of water, clear as crystal, over which, when you bend down in their classic depths, you can see your face." The conceit of the mirror, conjuring Narcissus, continued as Bartlett recounted the happy story of farmer's daughter Alice Neil, who "never did look sweeter" on a Sunday morning than when she appeared "dressed in white, with her little hymn book in hand" and who, when she raised her angelic voice in the church choir, made the old squires' eyes "moisten with love and tenderness." In Bartlett's short story, suitor Charles Davenport perceived in farm girl Alice Neil "beauty and grace and education . . . and, more than [that] love and kindness." When in the course of events Davenport asked his father for permission to woo the lovely country lass, the old man conceded, saying, "She may be humbly born, but she is nobler and far more worthy than those rich and fashionable women who live to ride in their carriages and look coldly down upon the virtuous poor!"

Such wishful projections continued apace after the Civil War, finding a wider and wider audience in the era of the nationally circulated periodicals such as *Harper's Bazaar*, where serial fictions singing the praises of country girls equal parts fatal allure and spiritual redemption became the norm. In the spring of 1872 popular Victorian novelist Mary Elizabeth Braddon published the episodic novel *To the Bitter End* in the pages of *Harper's*. The motifs were Romantic, focusing on the self-centered wishes of the leading man to leave the city "noisy with traffic" for the life of the "yellowing corn-fields studded with gandy field flowers, and the rapturous music of the lark, invisible in the empyrean." In the newsprint pages of Braddon's novelette, protagonist Herbert Walgrave could be found lamenting, "Oh, to be a country squire again, and to live my own life, to marry Grace Redmayne, and dawdle away my harmless days riding around my estate; to superintend the felling of a tree or the leveling of a hedge, to lie stretched on the grass with my head on my wife's lap, my cigar-case and a bottle of claret on the rustic table beside me; to have the renown that goes with a good old name and a handsome income." In Walgrave's arcadian vision the allegorically named Grace Redmayne is granted centrality as a vehicle to a more blessed bucolic existence. That he should love so simply strikes the cosmopolitan Walgrave as a mystery most worthy. "Strange that a farmer's daughter, edu-cated at a provincial boarding school, should exercise more influence over me than any woman I ever met—should seem to me clever and brighter than the brightest I ever encounter in society," the youthful Walgrave wondered aloud. For him, Grace, by her rural birthright alone, stands as something of a rebuke to a crass industrial system. Consequently, to court her in lieu of her city brethren was to symbolically reject the civilized life—a stretch for a man like Walgrave, who we are told had never "swerved by a hair-breadth" from the intention of working and marrying in the best, most urbane cir-cles. Still, swerve Walgrave did, the image of Grace Redmayne's many rari-ties sufficient to beguile him away from the city.

By the late 1800s the best kind of farm girls, the literature implied, were not to be found knee-deep in manure, nor tanned and leathered by real work under an unforgiving sun, but possessed of a countenance almost virginally white, saved from beauty-reducing toil. Circa the Chicago World's Fair of 1893, anxieties suffered over the "new" female body of an industrial age peaked in keeping with the Victorian notion then en vogue that the rigors of men's work would somehow "unsex" the fairer sex. For the World's Columbian Exposition, castings of a so-called average American boy and girl designed by Boston sculptor Samuel Kitson were constructed based on the "actual and precise proportions of the average human figure in the period of youth." Dr. D. A. Sargent had measured the students at Harvard University and other eastern women's colleges to create these representative, "average" figures. A writer identifying himself as T. W. H. wrote into *Harper's Bazaar*, arguing that the sculptures would put to rest any notion that intense study and physical activity risked altering young women's basic physiology—a farm-born notion that had long been used by Yankee cultivators as grounds to withhold rural maidens from the dirty life. So pale had grown many farmer's daughters as a result of this coddling, T. W. H. observed, that "the farmer's daughter [was] too apt to be pale and dyspeptic." Her body, he warned, had sufficiently atrophied as to have skewed the representative data used by the sculptors.

The World's Columbia Exposition, featuring the many electrifications, prettifications, and sophistications of the White City, marked the quickening of the great rural-to-urban migration that would, fifteen years hence, result in the formation of President Teddy Roosevelt's Country Life Commission. For generations many American rural girls, especially in the East, had moved from farm to city for boarding school or college, and by the late 1890s, the graver public policy concern was that boys would soon follow. Perhaps the first and most popular media magnate mobilized in the fight to keep American farm boys morally strong and thus resistant to the allures of the cosmopolis was the founder of *Wallace's Farmer*, "Uncle Henry" Wallace. Wallace had grown up on a farm in Pennsylvania before moving to the greener pastures of central Iowa to preach as a Presbyterian minister and to plow as a gentleman farmer. In his weekly "Sabbath School Lesson," he combined twin talents for agriculture and proselytizing into advice columns written for farm boys and their parents, pieces collected in his 1897 book *Letters to the Farm Boy* and issued by Macmillan in a remarkable five editions between 1900 and 1918. Wallace had earned his avuncular moniker by virtue of his homespun style, as evidenced by this, his entrée into the difficult subject of finding the right girl: "I sometimes think that it is essential to the right development of a boy that he should have, first, a dog; second a chum; and third and last, his best gal." In a letter titled "The Farm Boy and His Start in Life," Wallace cautioned his farm son brethren against dating "some good-looking girl, who . . . does not value you above one of her hairpins, who eats your caramels and ice cream, thinking, if she thinks about you at all, that you are a silly goose for wasting your substance in that kind of entertainment. . . . She knows you are wearing more stylish clothes than you can afford, and she secretly makes up her mind that while she will have all the fun she can with you, she will say 'Yes' to an entirely different sort of fellow."

While Wallace advised his readers to write-off fancy girls bound for town, Midwest farm wives began to clamor en masse for attention to be paid their humble daughters, whom, they feared, might leave for regional metropolises like Chicago, St. Louis, or Indianapolis unless their needs were soon addressed. To that end a June 11, 1896 edition of the *Hawarden Independent* in Hawarden, Iowa, ran a column under the heading "Let the Dear Girl Hunt Hens Rather Than Husband," penned by a woman identifying herself only as a "Rural New Yorker." "Much has been said and written on 'keeping the boys on the farm,'" the commentary opened, running alongside such curiosities as "About the Chicken Lice" and "Among the Poultry." The letter-writer continued, "Have you ever seen anything written in regard to 'keeping the girls on the farm?' Or are they of so little account that the only thought is to get them married off as quickly as possible so that the other fellow can support them, and so save that item of expense? I say to farmers, give your girls a chance, and they will beat the boys all hollow, not only in the thoroughness of their work, but in their enthusiasm for results, and in the results themselves."

By 1915 the nascent Country Life Movement had ebbed as Europe plunged into the Great War. Martha Foote Crow's foreword to *The American Country Girl* seemed to concede in advance the inexorable drift cityward of the nation's rural daughters: "By that time there will be seven million [country girls]—unless in fact these six millions shall have run away to build their homes and rear their children in the hot, stuffy, unsocialized atmosphere of town, leaving the happy gardens without the joyous voices of children, the fields without sturdy boys to work them, the farm homes without capable young women to, shall I say, man them?" While Crow's monograph lent country girls something more than the drive-by treatment afforded them by male authors, her book, marketed as much to urban bureaucrats and former farm girls living in the city as to those living their working lives on green acres, likewise suffered under the weight of its own ironies. Crow herself had long been a resident of Chicago and found herself more than fifty years removed from her own rural upbringing. Thus her treatment of the country girl is at times romantic, redolent of a similar whimsy deployed by male writers of the time. "It is the country girl that interests us," she waxed, "the promise and hope of her dawn, the delicate, swiftly changing years of her growth, the miracle of her blossoming. There is something about the kaleidoscope of her moods and the inconsistencies of her biography that fascinates us. The moment when she awakes, when the sparkle begins to show in her eyes. . . . We ask no greater happiness than once or twice to catch a glimpse."

JUST TWO YEARS after my teenager mother landed her first paying job off the farm in 1960, the evidence had already begun to mount that farm daughterhood was soon to change. On May 31 of that year the *Pioneer Press and Stanwood Herald* arrived at my paternal grandparents' house signaling far-reaching regional and familial shifts. On the front page ran news that the family patriarch, my farmer great-grandfather Everette Puffer, had passed away just one day after my grandmother's 45th birthday. Alongside his front-page obituary ran photos showing Mechanicsville youth swallowing a "Tasty Polio Vaccine." In the very same edition's "McVille Musings," columnist Pat Trump reported that 5,021 of Cedar County's 5,930 households owned a television set.

That same week's edition brought news of still more sweeping agricultural changes, including notice of livestock marketing meetings as Midwest farmers began to move inexorably away from dairy and beef, and imports of those same commodities began to rise. Cultivators without help at home were willing to try any labor-saving device that would replace the work of the ever-rarer hired man and the increasingly busy or absent farm children. The *Pioneer Press* ran a photo captioned "Trial by Fire," which showed local farmer Herb Crock attempting to eliminate weeds in his 18-acre field with the use of an "LP gas flame thrower." On the fertility side, Nies Grain and Feed, whose homey motto read "Where the coffee pot is always on," made a pitch for something called Arcadian Golden Uran liquid nitrogen solution. "Pumps and machinery do most of the work," the Nies ad promised, adding, "no pressure, no odor, no loss of nitrogen to the air."

"A financially sustainable commodity farm now has got to be 1,000 acres or more, but most baby boomer girls didn't grow up on farms of such size," farmer's son Kevin Woods tells me of the cultural and agricultural changes he witnessed in these, his teen years. "In the '60s a huge farm was 700 or 800 acres. The evolution of most of the 'family farms' of the '60s and '70s was either that they got much larger or they disappeared into larger farms. Most women I know who grew up on farms did just what most of us boys did—recognized that there was not enough farm or money to go around, went to college, got a 'town' job, or learned a trade."

A closer look at the numbers confirms Woods's recollection of a generational drift to greener pastures. In 1945, a decade before Woods was born, USDA figures showed just over a quarter of farm households earned off-the-farm-income. By 1970, one year before my sister entered the world as the last of seven consecutive generations of farm daughters, off-the-farm income had increased twofold to over 50 percent. And in 2002, just after my nephew was born—the son of a computer animation professor in New York—the number had increased to 93 percent. The cold statistics proved hard to argue with: if a farmer's daughter earned her appellation and cultural calling card by being a product of a self-sustaining farm capable of supporting a family, and if, as the old saying goes, you can't take the farm out of the girl, then the precipitous decline in Midwest farms meant the inevitable loss of the girl.

Woods, an engineering supervisor at Schneider Electric in Cedar Rapids, Iowa, still lives on the river-bottom farm where he grew up. Having sunk his roots deep in his midwestern farm community, he can think of only "a couple" of the Midwest farmer's daughters in his class of 1973 that remain on the farm, and for him their absence is self-evident. For Woods the story of what prompted the Midwest farmer's daughter to abandon her pedestal is economic as well as cultural, and he offers a characteristically wry recounting of the bittersweet farm economics of the late 1960s when he helped his father on the family's 65-acre row-crop and dairy spread. "I was forever thankful for whatever caused milk prices to collapse in '67 or '68, resulting in my dad selling our seven wretched, fence-jumping, tail-swatting, balky milk cows and allowing me to sleep in," Woods quips. But beneath the dark humor lies a deeper commentary about how sharply divided sex roles gave the region's farm daughters and sons plenty to rebel against in the already turbulent 1960s. "I don't recall that there was ever a formal conversation about the division of labor and rationale behind tasks assigned," he recalls, "but it was clear to us kids that Dad was Management and we were Labor, the fruits of that labor being evidenced in roof over head, food in belly, and shoes on feet. But we did know, without being told or engaging in a lot of angst-ridden navel-gazing, that we were an important Part of Something, something vital to the success or failure of our farm, and family."

Woods describes his parents' agrarian modus operandi without mincing words. "My mom's primary job was the care and feeding of four kids, stair-stepped twenty months apart in good Catholic fashion, keeping the house relatively clean, and being Worrier-in-Chief," he remembers. "Dad went to the bank to borrow another stake each spring in order to engage in another year of crop roulette. 'Family farm' unfortunately comes with 'family,' so Mom dealt daily with my grandfather who should have had a PhD in Taciturn. . . . She was also the bookkeeper and kept all the farm records in spiral bound notebooks and shoeboxes full of receipts."

"Our family dynamics were pretty traditional for a 1960s and '70s Midwest farm," he continues, "equal parts love, contempt, manipulation, and dysfunction, with my dad in charge and my mom able to artfully work him like clay as needed. Dad didn't rule with an iron hand, but you could certainly be on the receiving end of same if one were a mouthy 15-year-old boy showing great disrespect to one's mother. Leaving a waterer on all night or wrapping a disk up in a fence due to daydreaming could earn one a one-hundred-decibel counseling session, but I never doubted that he loved me, though he never said so until a couple days before he died."

Of the family's four kids, reports Woods, it was his sister who, ironically, showed the most enthusiasm for farmwork, despite being "chained to the house" by a father who preferred her to learn the traditional domestic skills she despised. "My dad was for a long time at a complete loss as to how to relate to a daughter who may as well have been a Martian. My sister chafed at the reins and broke free to be outside doing farm stuff whenever possible. Today she is hobby farming 40 acres off the home place with her husband, raising cattle, and three sons. She drives tractors, works cattle, grinds feed, fixes fence, and loses money. What could be more farmer than that?"

When I ask Woods what Corn Belt culture stands to lose as it loses its daughters of working farms, I expect an ironic answer—standard fare here when the specter of the farmer's daughter is raised, the concept so suffused with a culture's best and worse agrarian baggage that many tread carefully around it still. Not Woods: "Well, the Midwest loses young men, who, by and large, are greatly interested in young women, and if a young man has a fancy to farm, and isn't totally being led around by his hormones, he'll be looking for a young woman who understands what he wants to do with his life and wishes to partner in that endeavor. I hate to sound like a traditionalist fossil, but at their best women are the glue that hold families together, especially farm families. The men are too busy fighting the often losing battle with the weather, pests, bank notes, and dozens of things beyond their control. Farmer's daughters often have married farmer's sons, and as an extension of the home, have often been the backbone of rural communities."

When Woods and others speak of the farm daughter they invariably and involuntarily slip into the past tense, as if the very women they're referring to have become as anachronistic as a blacksmith's daughter or a cobbler's son. As far as Woods is concerned, a young, rural woman who cooks, cleans, washes, mends, reaps, sews, and herds—in other words, chores her grandmother likely completed as a matter of course—today "covers [few] outside the Amish community." In the fewer, bigger farms of the Google era, he maintains, the farmer's daughter is more likely to involve herself in the management or business end of the farm than the "backbreaking part" of the labor. "A delicate female hand can drive a tractor or combine as well as anyone," he asserts, "maybe better than many of us ADD males." Another place that the classic farmer's daughter may be thriving, he contends, is in the exploding community-supported agriculture scene, where "buy local" farmer's market and organic communities are perceived as attractive to urbane young women with agrarian sympathies.

"Much of what I have said here are generalizations about farmers' daughters, dangerous waters to be sure," Woods cautions in closing. "I think farmer's daughters grow up quicker in some ways, just like farm boys, because life and death are in front of them every day. Life isn't fair, food looks back, and toddlers on a farm learn soon the tragic nature of life, that it invariably ends badly. I find the farmers' daughters I know, who range from forty to eighty, are feminine, and sometimes beautiful. But there is a toughness there, too, and folks that trifle with them do so at their peril."

Five years behind my mom at Mount Vernon High School, Woods, as he hit puberty, began to covet the attentions of the last real-life farmers' daughters who worked the fields near his hometown. Images of the farm girls he admired outside his window infused his dreams, becoming his first real adolescent crushes. He recalls:

> I remember Debby M., who lived on a neighboring farm, a small enterprise down by the river. . . . She was the oldest of six kids, the rest boys, and unruly BB gun-toting, rock-throwing, eye-punching, trap-line-running hooligans they were, too. . . . She was very sweet, and was my first real crush at about twelve or thirteen. We all rode the school bus, and since we were about the last on the route, it was sometimes Debby and I alone, last to be let off. I agonized for literally months before I mustered up the courage to sit by her, sharing the green seat on the big yellow Blue Bird limo for a few miles until they let her off. The bus driver, Al, would look up into the mirror over his head, used normally to scowl kids back into their seats, and he would smile at me and wink. Debby and I never said a word to each other while we sat next to one another, but would chatter away otherwise when at school or around the neighborhood. Not sure why. Maybe the big green bench seat on the bus was just too intimate for junior high. Her dad died and they moved to town. I saw her at a high school reunion a couple decades ago, but she didn't have much to say. Evidently, everything had been said on the Blue Bird Express.

I remember Sandy K., who was my age and lived a mile or so away and was gorgeous when she was young, who raked hay on an old John Deere 630 wearing jean shorts and a bikini top, working on her tan, no doubt, but nearly causing some of us to drive our motorcycles off the gravel road as we gawked unashamedly and tried to think of reasons to drive by during haying time, or better yet, get her old man to hire you to bale. I heard news of her, some forty years later, that she had been divorced twice, been through treatment for alcohol abuse and is now clean and sober, but that life had more or less beat the hell out of her. I prefer to remember the smiling teenage girl raking hay on sunny summer days, who no doubt marveled at all the traffic on the gravel road during hay-making time.

FOUR DECADES AFTER Brian Wilson crooned the iconic virtues of the region's farm-raised females, "California Girls" has been paid frequent homage, from the Beatles's "Back in the U.S.S.R.," to the 1985 cover by Van Halen's David Lee Roth, when the tune once more cracked the top five on the Billboard charts as it had three decades earlier when my mom listened to Billy Joe Royal's plea for a boy from down in the boondocks. Once more in 1985, a farmer's daughter and son—my sister and I—found ourselves singing along to the by-now intergenerational theme song.

Nearly fifty years after the Beach Boys bested *The Sound of Music* soundtrack that told the tale of a real rather than mythical country girl in songs like "The Lonely Goatherd," the phraseology the Wilson brothers popularized continues to serve as a semantic reference point for those removed from the both the farm and the region. "OK, so I'm really a Midwest Farmer's Granddaughter," a blog calling itself the "Midwest Farmer's Daughter" begins, "but I didn't want to disrespect the Beach Boys by screwing up their lyric. The fact remains that I'm a simple Midwestern girl living in a power-crazed East Coast world." Meanwhile, a popular Internet travel site, NileGuide, deploys the timeless turn of phrase to drive tourist traffic to the forgotten middle of the country, opining, "If you're searching for that 'Midwest farmer's daughter' the Beach Boys sang about, there's probably no better place to find her than in Indiana. From overhead, Indiana looks like a great patchwork quilt of green, but down on the ground, there's more than just rows of crops and barns."

Meanwhile, the Midwest farmer's daughter has become a national rather than merely regional identity, a calling card embraced by the granddaughters and great-granddaughters of the country girls who once literally put bread on the table for a hungry and growing nation. Like a host of terms once considered circumstantially damning—handles like *working class* or *blue-collar*—in certain circles the term itself has been reappropriated as a badge not of objectification, but of an exceptional heritage hard-won.

I get in touch with an acquaintance of mine, Becky Kreutner, a teacher in the Iowa City metropolitan area, to separate fact from fiction, wheat from chaff, regarding what today's country Jill feels about her rural heritage, and how she may feel about continuing it. Kreutner grew up on a 200-acre farm in aptly named Eden Township in Benton County, Iowa, and more than any farmer's daughter I know, she embraces her birthright.

"I definitely dreamed of marrying a farmer," Kreutner confesses, "who would most likely be a farmer's son. I loved living in the country and envisioned myself having a life very similar to my mother's and my grandmother's, in which I would be farming the land together with my husband. However, I always dreamed of being a teacher as well. After I accomplished this dream and moved to the 'city,' I realized that that there was a whole different way of life. One that still involved hard work, but more mental work than physical labor. One that wasn't dependent on the weather and the markets, things that you had so little control over. I have to admit I enjoy this type of lifestyle a great deal. Although I still dream of marrying a farmer's son who shares my values and upbringing, I no longer envision myself living on a working farm."

Audra Brown, meanwhile, grew up rumpusing around her family's 100-acre farm in Louisa County, Iowa, on the banks of Long Creek. "I always assumed I would marry a farmer's son," she replies when I ask her what she imagined the future would hold in those halcyon days, "and I dated one for most of high school. That certainly changed when I attended college at the University of Iowa. I guess you could say I was 'fresh off the farm,' and my curiosity about other people and cultures drove me to try things I'd never experienced before. The boys back home held a very narrow definition of beauty (blond, busty, and blue-eyed), and I despised that (probably because I was a freckly, flat-chested, short-haired brunette.) The excitement of the city life appealed to me, and I dated men of all different ethnic backgrounds. I ended up marrying an African American man from Boston."

Eventually, I return to Woods for his sober perspective on the romantic question that has occupied American songwriters and poets for two centuries—what's in the Midwest farm daughter's soul, and what weighs most heavily on her heart. Woods admits he doesn't know. "My guess is that she struggles with the fact that a way of life that consists of small farms, small towns, and community is dying," he confides, "and will not come this way again, at least not in her lifetime, unless petroleum dries up tomorrow, as modern food is every bit a petroleum product just like plastic. Agriculture is simply a wall-less factory and will remain so."

Woods's comments suggest that even if the latter-day Midwest farmer's daughter now living in the city returns to her acres after achieving financial stability, her renaissance will not arrive soon enough to turn back the clock, nor in time to replace her parents' generation on the farm as en masse they step away from the combine and the cattle call. All of which means the *daughter* in the iconic phrase *farmer's daughter* may itself be poised for an early retirement to a place where neither song nor sentiment may reach, lingering there, just beyond the tip of the tongue.

CHAPTER THREE

HOW YA GONNA KEEP 'EM DOWN ON THE FARM?

"**IT WAS A** time of disillusionment. Three heroes had been lost—Jack Kennedy, his brother Bob, and the Reverend Martin Luther King. As my grandmother would put it, 'I wanted to fly the coop.'"

Dianne Ott Whealy, the girl who wanted to fly the coop, is cofounder of Seed Saver's Exchange in Decorah, Iowa, an 890-acre farm and seed exchange whose sales, in 2008 alone, were estimated at $2.4 million. In fact, farm daughter Ott Whealy, whose celebrated acres hosted Barack Obama on his 2011 rural tour, may one day be responsible for saving life as we know it. In 2008 Whealy's Seed Savers Exchange was part of the first shipment of one hundred million seeds to be stored in the Svalbard Global Seed Vault, located four hundred feet deep in a frozen mountain, near the village of Longyearbyen on the Norwegian island of Spitsbergen above the Arctic Circle. "Svalbard is the lock box, the safety deposit box, for the world's seeds," Drake University's Neil Hamilton, vice chairman of the Exchange, told the *Des Moines Register's* Jerry Perkins on the occasion of the deposit. "This is the place that, when we need to go there for seeds, they will be there."

However, Ott Whealy confesses why, in 1968, a farmer's daughter was the last thing she ever wanted to be. At that time, eager to transcend what then seemed to her a homely past, she and four other teenage girlfriends boarded a Midwest train and headed West.

"It was the beginning of all the flower children and protests and anti-establishment, and I just fell right into that," Ott Whealy tells me, "and I was anti-my life too at that time. It was a time when I thought, 'Poor me. I grew up on a farm in Festina, Iowa. How corny is that?' I remember not even admitting that fact. When people would ask where I was from, I would say 'Cedar Falls' because I was embarrassed I grew up on a farm. And then that gradually shifted back as you go through the years. . . . I think I had a lot to rebel against . . . including my parents . . . including my life." Looking back, Ott Whealy remains somewhat incredulous that her farm-girl self could have been so naive. "Here I am, just growing up in this little corner of the world. . . . I just thought that there wasn't anything more than this, and then I realized that there was. I would definitely say I was restless, you know, at that time in my life, and I went off and totally went crazy."

Sometime in the era of Nixon, somewhere in the middle of the Midwest, a genre was born—the bittersweet memoir of the farmer's daughter who left the farm unapologetically in the era of the Bobby Kennedy and MLK and Tricky Dick, and settled down in the city among a surfeit of books and movies and an endless tableau of new, educated faces, and looked back, if at all, in memoir. She left home for the nearest metropolis; she lives there still, in fact, in some college town two hours or twelve from her family's home place, and her heart is big and her understanding of people and animals and human nature is deeper than most of her peers, and she sometimes yearns for a life she played a large and often unwitting part in killing. In other words, Ott Whealy, one of the most visionary and successful women entrepreneurs of her era, is far from alone.

By 1969 Ott Whealy had returned to the Midwest from her "time of exploration" in Estes Park, Colorado, where she roomed with her girlfriends and waitressed at a café, moonlighting at the Plantation Restaurant to try to make enough money to feed herself and her oversized Saint Bernard named Ulara. Her midwestern farm parents visited once, and they were horrified to learn their daughter couldn't direct them to her local Catholic church. Ott Whealy gave a year at Minnesota's Winona State University a college try, but ended up dropping out to return to Aspen with her friend Patti. "They didn't care at all," Ott Whealy recalls of her parents' reaction to her decision. "It wasn't like, 'You're not quitting! It was like, 'Oh, ok.' The expectation was for me probably to go to church every Sunday, get married, and be a farmer's wife." However, for her brother—the farmer's son—the expectations were different. "He didn't like farming. He always said he hated it, so he went off to school where he could do something else. My parents' farm was so small that he couldn't just farm with my dad anyway. He never had that interest in taking over the farm or even being there. I'm sure that played into it as well. I think that they were proud that he went off to school. Then I went off and partied."

Across the Corn Belt farmer's daughters were casting aside their mother's recipe books for what they increasingly saw as the good life lived elsewhere. In their grandmothers' era, popular books like William Arch McKeever's *Farm Boys and Girls* made a point of reminding country girls that the chestnut never falls far from the tree. "It is the mother that shapes and molds the character of the girl. If she is sweet-spirited . . . her daughters will as a rule have the same sort of outlook. . . . If she sharply criticizes the preacher's sermon at the Sabbath dinner, she need not expect her daughters to become devout. If she is a poor housekeeper, how can she expect her daughters to excel in that finest of arts?" McKeever asked in 1912. That same year, when my Great-grandmother Amber Jane Pickert wrote impassioned letters to friends expressing her intentions of becoming a farmer, McKeever spoke to the necessity of "carefully training the growing children to perform such deeds as will shield the mother in the home."

Given the generation gap then opening up between baby boomer children and their parents, shielding mothers hardly seemed possible, let alone advisable. A case in point is my mother's mother, a master at the very domestic arts about which my mother harbored the most profound doubts. "She was very accomplished," my mom recalls of her mother. "She was bright. She was very attractive. Yes, she was quite the catch, as was, on the other hand, my dad, who was at that time was very shy. Even back then our name was a symbol of wealth. Back then, too," mom continues, "the farmer was king. A farmer who did well and made money and had good crops and nice fields was really looked upon as a leader." The farmer's wife, especially if she was as resourceful and adept as my grandmother, was, by extension, a reigning queen.

"Every time I used the sewing machine I broke it," my mom recollects of her doomed attempt to live up to the angelic domestic standard set by her mother. "They didn't trust me at all with the riding lawnmower. Mom did not want me near anything mechanical." Indeed, beyond the menial work of washing dishes and setting the table, my mom was verboten in the kitchen, where her gifted cook of a mother found too many hands made for more work rather than less. "As farm girls she really didn't want us in the kitchen. Once a month we took turns baking," mom remembers. "But she did not teach us about cooking, nor did she really want us in the kitchen." In McKeever's book the daughter's duty to "shield" the mother had been based on common sympathies, but if those affinities had never been built, cultivated, or culturally supported, what then? It wasn't long before my mother's generation of farm girls began asking themselves who they were protecting, and at what cost.

Even the best-intentioned farm girls of the 1960s often and sometimes unwittingly upset the apple cart of their mothers' well-mannered courtship habits. While manuals like McKeever's once advised, "very good woman ought to be taught how to know a good man," and "the young girl on the farm should have much advice in respect to the nature and character of men," many baby boomer country girls preferred to live more dangerously, shopping for men that proved opposite to their distant, often stoical fathers. "I thought I foresaw, in one bright flash," memoirist Debra Marquart writes of her first college boyfriend in Fargo, North Dakota, "what would be my life: a fat-faced baby banging a spoon in a high chair; strained carrots on the linoleum; a sky-blue telephone on the wall, ringing incessantly; his mother, my mother, always on the other end of the line wanting, I don't know, to exchange recipes." Marquart reports that such a humdrum fate seemed the death of her wish to "be something—a social worker, a rock star—a revolutionary."

Whereas McKeever's turn-of-the-century advice book had warned farm girls against city boys, reminding his readers that "many beautiful and innocent young women are lead astray either before or after marriage by vain and designing men . . . who harbor within them some most serious and insurmountable evil and disease," it seemed that the '60s and '70s required an equal-and-opposite warning for city boys who might be ensnared by designing farm women. "This was no little house on the prairie," Marquart offers as her page-one introduction to the North Dakota country girls she ran with in the 1970s. "We smeared musky blue shadow on our eyelids and raspberry gloss on our lips. We wore bell-bottom jeans. . . . We were hip-huggered, and tight-sweatered, and navel exposed. We walked around town like the James gang, tossing this and flashing that."

My own mother looked to provoke a response from her parents, when at 16, she began dating a farm boy rebel—my father—who came from what was considered a "very good family" some half dozen miles distant. Still, the parents of baby boomer farm children found it difficult, in a changing world, to follow the same sage advice manuals that had decorated their parents' shelves. My mother's parents had not been, as McKeever advised, "for years purposefully engaged in preparing their daughter to know at sight a good man." Perhaps they trusted implicitly their daughter's choices. Or maybe they themselves felt sufficiently threatened by the culture wars of the 1960s to find solace only in the impossibly fresh-faced girl-boy duos showcased each week on *The Lawrence Welk Show.*

As a freshman at Mount Vernon High School, my mother had begun taking college-track courses at a time when, she says, her peers were either considered "college material" or they weren't. Still, when the time came to apply, she felt at sea. The university she proposed to attend was 90 miles away, and she didn't have a car, a driver's license, or parental support for her ambition to become an elementary schoolteacher. A generation earlier she might have practiced her intended trade at the local schoolhouse, but with the introduction of stringent, some said draconian, state certification requirements, her best hope now lay with the Iowa teacher's college hours away. At times my mom wondered whether her parents' disinterest might be a punishment for aspiring to something beyond the homemaker role they had envisioned for her. Good grades, which my mom and her sisters brought home in spades, seemed to attract her parents' interest only if they were lacking. "There was an expectation that we got good grades and that we were good people," my mom recalls. "Maybe it's because we didn't dare come home with bad grades or get in trouble."

A larger meta-narrative of those days emerges in the memoirs of Midwest farmers' daughters who grew into young women in the late 1960s and 1970s. Marquart, a professor of creative writing at Iowa State University, writes in *The Horizontal World* how coming of age on a farm outside Napoleon, North Dakota, she learned to abandon the slogans on the posters on her bedroom wall that read "Grow where you're planted" in favor of more timely mantras such as "Label Jars, Not People" and "Make Love, Not War." Like my mom, Marquart was spared by gender, birth order, and generational timing many of the chores on the farm performed by her mother and grandmother. To her father's amusement, she "drove" cattle—or rather they drove her—as she listened to "In-A-Gadda-Da Vida" and smoked cigarettes.

The late 1960s and early 1970s seemed for a time to kick the myth right out of the farmer's daughter. Maybe country girls had always, contrary to their literary reputation for chasteness, lost their virginity young or went to town with the wrong sort of men, but the innocence mantle no longer seemed theirs to have and to hold after the Beatles appeared on *The Ed Sullivan Show*. In Marquart's memoir the dissonance is sufficient to induce the third person in her lyrical yet matter-of-fact litany of things experienced:

> She's tried French kissing and seen plenty of window fog. She's had her breasts nibbled, studied, stroked. She's explored an endless variety of back seats, quietly undoing snaps and zippers and trick belt buckles, slipping her hand deep into the moist pants of some boy. The horizontal life has not eluded her.

> But she has impossible standards. She does not want the tentative touch of a novice, nor does she want a clumsy farm boy. No one who wears tidewater plaid plants, smells of cow manure, and lusts more for her father's full six quarters of land than her own strong, slim, tanned body.

> She's no longer interested in shy exploration, or wonderment, or handing her carefully pruned virginity over to a husband on a wedding night. She has an itch somewhere deep inside her, in a place she cannot even begin to direct someone to.

In nearby Minnesota, a young Gayla Marty was feeling the same pull away from her family farm outside Rush City. A dutiful Baptist daughter, she recounts in her memoir, *Memory of Trees: A Daughter's Story of a Family Farm*, how she commenced in high school taking as many shifts as she could at the local restaurant, Grant's, to escape the drollness of the farm; how after the café closed she and her girlfriends would cruise the nocturnal country roads listening to Grand Funk Railroad, Foghat, Led Zeppelin, and Deep Purple. In school a sophomore boy committed suicide after getting a girl pregnant, and Marty writes, "My simple assumption that I would go to college and marry and make a family ran into the river of Mama's sorrow about her own life." Suddenly, it seemed, "unhappiness descended anyway, deeper than I had ever felt, sharper, and more painful, like stones tied to arms, legs, and neck." In the next scene Marty temporarily loses heart and, after driving home after a party, stops on the railroad tracks, flirting with suicide.

The lifestyle of the farmer's daughter seemed suddenly out of step with the hippie movement, at least in the eyes of the girls who had lived the rural life rather than merely read about it. The rural Midwest often appeared to her, much as it had to Hemingway and Fitzgerald and the rest of the Lost Generation, grim and lifeless. In 1973 Minnesota writer Carol Bly voiced these and other seemingly heretical thoughts in a series of thirty-one essays produced for Minnesota Public Radio and eventually published by Harper & Row as *Letters from the Country*. At the time Bly had moved back to her hometown of Madison, Minnesota, just as Diane Ott Whealy had a few years earlier moved back to her own home place. In opinion pieces Bly found herself wrestling with the identity of the place she soon dubbed, after the fashion of Fitzgerald, her "Lost Swede Town." "When I came out here I thought it was just sexual loss," Bly wrote in September of 1973. "On my first visit, we drove in the evening. The bare bulbs were lighted in the passing farmyards. I remember saying, How marvelous to think of night on this gigantic prairie—all the men and women making love in their safe houses guarded by gloomy groves! . . . The reply was: That's what *you* think!"

In time Bly began to see what had been meant by her traveling partner's cynical reply that night, registering what she called the "sexual chill" of her Scandinavian American climes, struggling to name a feeling she came to characterize as "a restraint against feeling in general" and a "restraint against enthusiasm." Nobody, Bly lamented that first fall, wanted to help her Lost Swede Town wake up. "The twentieth-century way to look at the nonverbal, non passionate Midwesterner," she opined, "[was] to sneer." As her monthly commentaries grew in number, Bly became increasingly engaged in what she thought was constructive criticism of her midwestern culture and its seemingly native oppressions, though many of her fellow residents did not much truck with what seemed to them hyper critical and mean-spirited commentary. "Our countryside has inherited not Grieg, not Ibsen, not Rölvaag," Bly chastised her countrymen, "but just sitting there, cute movies, and when the boredom gets bad enough, joining the Navy." Arriving in Madison, Minnesota, with a film crew years later, years of Bly's well-intentioned goading had rubbed nerves sufficiently raw that a man at her local VFW laid a threatening hand on her photographer and director and said, "You are that goddamned TV crew working with that goddamned Carol Bly."

For natives who had grown up in such cultures—natives like my mom, Bly, and Ott Whealy, homegrown traditionalism sometimes stifled and repelled. In her 2011 memoir *Gathering*, Ott Whealy remembers returning to her small Catholic town with her out-of-state fiancé Kent, who, though he had short hair and a college degree, turned out to be of the wrong religious persuasion. Father Graff at the Dolors Catholic Church in Festina would only agree to marry off his hometown farmer's daughter if she had exhausted all the other eligible Catholic men in the community (she hadn't), become pregnant (she hadn't), or promised to convert her beloved to Catholicism (she hadn't, but she pledged she would).

East of the Lutheran and Catholic climes of northern Iowa and Minnesota, Janice Hill's blond-haired, blue-eyed mother was experiencing an even more profound disillusionment with her native Midwest. Hill, a community planner in Geneva, Illinois, known across the Midwest for her county's progressive, pro-sustainable farming policies, confesses that her mother left the family's Michigan farm at 14 to live with a friend, and accelerated her high school graduation date by two years so she could leave for the city.

More than fifty years later Janice Hill tells me, "My situation was the most extreme that could be, in that first of all my dad is African American and my mom is white from German descent." Hill insists that, for her parents, interracial marriage in Indiana in 1959 would have been unthinkable if not, she suspects, illegal. "My mother wanted to leave the farm because she did not want to be a farmer's wife," she says. "She told me she looked around at the farmers, and she wasn't interested, and that was it for her." Hill's mother ended up traveling to Chicago on a scholarship to a now defunct Baptist missionary training school. Instead of returning to her parents' Michigan homestead in the summer, Hill's mother stayed in the city to work for Carl Fischer Music, and one day her dreamboat walked in. "They met and fell in love. Somebody saw them in downtown Chicago from her small town, and said, 'I've got to tell my mom!' Then starts the journey for our family in terms of dealing with a mixed race couple."

"My grandparents had never been exposed to anything like that," Hill remembers. "When I came to rural Michigan as a baby, I was the first person of any color at all. . . . My mom was blond and blue-eyed, and so was the rest of her family." Years later, after a successful career as a teacher and principal in the Chicago Public School District, Hill's mother remained adamantly opposed to returning to a rural lifestyle in Michigan or anywhere else. "Even when my parents had enough money to buy, to invest, and my dad would ask, 'Should we by some land out there?' my mom would say, 'Absolutely not!' She had worked so hard to get away."

Just across the county line from the Chicago suburb where Janice Hill's mother later moved with her kids, lived young Anita Zurbrugg, a gifted Midwest farmer's daughter coming of age in a large and longstanding Irish German Catholic family in DeKalb County, Illinois. The current Midwest director of the American Farmland Trust, Zurbrugg grew up in a family numbering eighty-four cousins originating in the nine separate clans headed by her various aunts and uncles; someone in each, she tells me, still serves as a farmer or farm manager. Zurbrugg is the fifth of seven born-on-the farm kids, and she comes from an impressive family of four Illinois farm daughters that includes not just Anita, who with her law degree also heads the Center for Agriculture in the Environment at Northern Illinois University in DeKalb, but also her sister, geneticist and beef rancher Sara Faivre-Davis, who serves as a Barack Obama appointee to the board of the Federal Agricultural Mortgage Corporation.

Zurbrugg graduated from DeKalb High School in 1974, and she remembers with amused chagrin that the theme of her graduating class was

"Booze and broads and we want more. We're the class of '74!" Her parents
had intended for Anita to continue her education in the county's Catholic
schools, but with the closure of the local parochial high school prior to their
daughter's freshman year, a public education proved the only real alternative.

Entering high school in 1969, Anita recalls feeling "insulated" from
the highly charged political atmosphere her older siblings had participated
in at university. "I was certainly raised very much a Democrat and a progressive, but . . . our family never had a TV until I was 16. Not out of any
religious or philosophical reasons; it was just TV was a seen as a waste of
time," she recalls. As her older siblings came back to visit the farm in the
late 1960s, Zurbrugg recollects the "vivid discussions" that would ensue. "I
think I was probably a little naive at that stage and much more focused on
my life. On the other hand, I was very aware that one of my older brothers
was a housing association president at University of Illinois. He was very
involved, and could have gotten in trouble. And my oldest sister was very
involved in civil rights issues. That was a bigger issue perhaps than the war.
So I was raised with that awareness, but not the involvement."

Zurbrugg considers hers a "transitional generation" wherein she and
her fellow rural female contemporaries did not necessarily think of themselves as future breadwinners. "It was still a conventional outlook," she insists, "where either you were going to be a teacher, a secretary, a nurse, or
something more conventional. I look at my daughters, and there's really no
difference in how they or their husbands look at life and what they might
be doing professionally. . . . Now I look back, and I'm saying to myself,
'Right now I could have very much been the partner in the family farm.'
Had I done things over again, maybe I would do things differently. But at
the time [farming] was not a consideration. My education was really to fall
back on. The plan was we were going to start a family, and when the kids
were older, then my goal was to have an orchard."

Zurbrugg remembers her father as unusually even-handed in his treatment of his own farm girls and boys, and yet it was her beloved late husband,
Mike, her three brothers, and her father who ended up farming together
as a limited liability corporation, even though Anita had more direct experience with farming than her husband did. "I taught my husband how to
drive a tractor," she recalls, adding, "but my father had a pretty traditional

attitude and that expectation [to farm] was never created. So it's like the lightbulb never went on in my head that, wow, I could be farming, which I find now a phenomenal disconnect."

MY OWN MOTHER experienced a similar disconnect in the late 1970s. She had moved her family into a rented four-square farmhouse just a half mile down the road from her in-laws, who gladly took their daughter-in-law and her two kids under their wing as they had so many young people. "She did not do outside work, and wasn't expected to," my mom recalls of the older farmer's daughter that became her mentor, my grandmother Julia. "She had her role, and [her husband] had his. The thing that was different was that she embraced me. I loved what she did for me. She never judged. She just embraced me as one of the family. I loved her for that. When you kids got a little bit older, she introduced me to the woods, and taking all the grandchildren down and building a fire. Even though I had gardening in my background, she introduced more. She absolutely loved the country."

Still, the frustration of rural living that my mom had known as a girl had begun to reemerge in the early days of her marriage, and, though she was native to farm country, her reactions to the growing isolation she felt sometimes conformed to the feelings of displacement reported by many of estimated 1.6 million back-to-the-landers who had moved, unprepared, to the nation's hinterlands in the early 1970s. Mom recalls of her mother-in-law, "She was also, like myself in later life, a free spirit, and sometimes the confines of the rural life were not always in our best interests or really where our heart might be at times."

In 1979 my mother temporarily separated from my father, pulling up stakes for an empty farmhouse her farmer father owned some half dozen miles away. She was still living in the country, but this time she was living by her own dictates. Mom's story was an increasingly familiar one. From 1965 to 1975 the divorce rate in America doubled, with steeper break-up rates among back-to-the-landers, according to Eleanor Agnew, author of *Back from the Land*. "Given the epidemic divorce rate, break-ups were not surprising, really, merely ironic," she observes. "We had moved to the land expecting a kindler, gentler place, a nurturing space where we could spend quality time with our partner shelling peas and canning tomatoes." Instead, the farm daughters and back-to-the-landers of the 1970s seemed taken aback by what their parents or grandparents might have told them—that no struggle was more laborious, and at the same time more worthy, than working the land in partnership.

Baby boomers proved to be a wholly different generation—more self-aware and media-savvy, more mobile, and more wealthy than any generation the world had known. While agricultural community mores had discouraged many of their poorly matched parents from disunion, Agnew writes that divorce in the popular culture of the early 1970s had been made to look "downright chic," and a "tool for personal growth." Many back-to-the-landers felt out of their element in the rural landscape. Their vision had been purely idealistic, removed from the practical experience with the agrarian skills and attitudes their new life required. "Like most of my peers across the nation who sought genteel poverty on the land, I too had grown up in a middle-class family, insulated from any conspicuous connections between work, money, power, and class," Agnew reflects. "My father disappeared into the sliding doors of the commuter train each morning and reappeared hours later at the dinner table. Any relation between his daily rides into the city and my safe, clean, comfortable surroundings was all but lost on me."

While it would have been reasonable to expect farm girls like my mother and her sisters to be more adept at rural life than their suburban peers, changing demographics and economics on the farm suggested otherwise. In fact, the circumstances faced by my rural mom and aunts proved strangely similar to those recounted by the urban Agnew. Though their farm father did not carry a briefcase and disappear into a train, he would be hard at work in the fields or dairy barns by the time the school bus arrived early in the morning to ferry the girls to town. But unlike Agnew and her citified peers, my mom and aunts could not expect to see their father at six o'clock sharp for dinner, especially when weather compelled extra-long hours in the field. A generation of farm daughters, and daughters more generally, had grown up not witnessing mom and dad working together, side by side, as the pioneer stereotype implied, but playing out strictly gendered divisions of labor.

During the Reagan era the idea of highly educated, highly paid, highly intentional couples enduring a true homestead experience, let alone a sink-or-swim immersion on an actual working farm, seemed to many cultural observers less and less likely. "Eventually we came back from the land," Agnew writes. "Either we came back literally, by leaving the farm and moving to an urban area, or we came back figuratively, by staying on the land but slowly taking on a way of life befitting middle-class people who just happen to live in a rural area." Agnew herself left urban Massachusetts with her husband and young child in 1975 to build a rustic cabin in the woods at time when inflation had reached 14 percent and the country was still recovering from the oil embargo of 1973. Less than four years later, Agnew had left the cabin and her husband, divorcing in 1981. Her marital experience, and those of many back-to-the-landers, followed a pattern closely analogous to the farm daughters of the era, many of whom had never intended to stay on the land or to earn their livelihood from it. "In most cases the 1970s back-to-the-land story can be told collectively," Agnew observes in the preface to her book. "A person goes to the land to be self-sufficient and free, the freedom loses its luster when the poverty grinds, the person and his or her spouse divorce, and the person slides back into the mainstream, gets a professional job or entrepreneurial gig, and remarries."

By the early 2000s new "asymmetries" in the urban "marriage market," reported Lena Edlund in her provocative study "Sex and the City," meant that mobile young men and women were increasingly likely to delay marriage, and to meet better-educated, higher-earning partners in the city, embodying a trend toward what social psychologists called "assortative mating." In a newly mobile world a young person, including the iconic farmer's daughter or son, could vote with her or his feet for the places that reinforced aspirational professional and personal identities, author Bill Bishop noted in his book *The Big Sort*, places that, unlike the stereotypical farm, offered the solace of living in close proximity to the educationally and socially ambitious. "These weren't political choices," notes a highly circumspect Bishop, "but they had political consequences." One of those consequences, some cultural critics felt, was a generation of farmers' daughters and sons, yet unborn, whose would-be rural heritage had been sacrificed to assortative mating and, quite possibly, economic necessity.

"At least once each day the busy farm father may think of a way to combine his work with the children's play."—From Farm Boys and Girls *(1912)*

"Sowing the seed, all by herself."—From Farm Boys and Girls *(1912)*

CHAPTER FOUR

RAISING
FARMER JANE

THE JULY 8, 2007 edition of the *New York Times Book Review* led with a blast from the past: the life of the farm daughter had once more become fodder for the best-seller list. For eight solid weeks Barbara Kingsolver's tale of the recovery and miracle wrought in her own daughter by Kingsolver's mid-life move to the farm had lodged itself in a list that included such heavy hitters as *The Diana Chronicles* and *The Reagan Diaries*. Still, the book truly turning heads that summer was *Little Heathens*, picked as an Editor's Choice by virtue of its author's "soaring love for her childhood memories" that coaxed the reader into "joy, wonder, and even envy." The *Christian Science Monitor* jumped on the bandwagon, naming Mildred Armstrong Kalish's recounting of a midwestern farm girlhood among the "books we liked best."

What most knocked urban critics off their concrete tuffets wasn't so much the book's setting—the Depression-era Midwest—or even its genre—

the seemingly shopworn memoir—but the simple fact that a Midwest farmer's daughter had once been so exquisitely happy with her lot. In the glowing *New York Times* review that cemented the book's ascendance, writer Elizabeth Gilbert opined of Kalish's laconic narrative, "No self-respecting modern memoirist (myself included!) would ever abandon such a juicy bit of suffering as a banished father. Surely one could milk volumes of pain (and book deals) from such misfortune!" A month after Gilbert registered incredulity at the story's simple pleasures and its refusal to cash in on the farm's more salacious goings-on, the *Christian Science Monitor*'s Yvonne Zipp jumped on the proverbial hay wagon, praising *Little Heathens* for giving off "an air as clean as laundry" fresh from a farm clothesline. "For those of us who have never brought in hay, sown potatoes, or killed our own dinner," wrote the reviewer, "the book will make you realize how easy life in the US has become." Picking up on the theme, Zipp's colleague Marilyn Gardner lumped both literary farm girl books into one, writing that Kingsolver's *Animal, Vegetable, Miracle* and Kalish's *Little Heathens* together proved that urban and suburban childhoods had become "pampered" by comparison. "Younger generations have little knowledge of where food comes from," she lamented.

Little Heathens managed to capitalize on, while simultaneously serving as an antidote to, pervasive anxieties about a generation of tuned-out, plugged-in children. The utility of a return to "the dirty life," as Kristin Kimball's 2010 memoir of the same name termed it, registered once more with a wide cross section of book-buying Americans. Particularly convincing had been Richard Louv's 2005 surprise bestseller *Last Child in the Woods: Saving Our Children from Nature-Deficit Disorder*. Louv, the author of a handful of books concerning nature, family, and community, and an adviser to the National Scientific Council on the Developing Child, quoted studies from the Robert Wood Johnson Foundation linking kids who lived an "indoor, sedentary childhood" to mental health problems, alongside a British study that found that the average 8-year-old was faster at identifying figures in the Japanese card game Pokémon than native plant species in his or her own home communities. Curing the nature malaise, a slew of subsequent studies seemed to suggest, meant cultivating more old-fashioned childhoods, throwbacks to the way we grew up way back when, on the farm. Researchers at the University of Illinois, for example, learned that exposure to nature may help reduce attention deficit disorders among children, while teams at Cornell University in 2003 concluded that children who merely grow up with a "green view" outside their window proved better equipped to handle stress.

For all America's intensifying urbanity, the desirability of the country estate was as ancient as Rome, and in America, as old as Jefferson, who considered the daughters and sons of agriculture the "chosen people of God." In the new millennium polls still showed that more than half of Americans believed farm life was more honest, moral, and virtuous; 64 percent, reported agricultural historian Doug Hurt, believed that farmers were harder working than average, and nearly 70 percent saw them as closer to their families. Moreover, by the early 2000s, one of the last and best arguments for suburban living—health—had come under fire in medical journals such as the *American Journal of Health Promotion*, where Reid Ewing and his team at the National Center for Smart Growth at Maryland University had found suburbanites across 488 sprawl areas to be, on average, 6 pounds heavier than their urban peers, while suffering marginally higher blood pressure.

But for all the supposed benefits of growing up on a working farm, fewer and fewer Americans—less than 2 percent—chose to live on a farm, let alone raise their children on a working commodity operation like the one on which Kalish had, in part, grown up. That the warmth with which the aged Kalish had remembered her uncomplicated rural childhood had struck readers as both exceptional and refreshing spoke volumes about changing views of farm childhoods from one generation to the next, and from one phase of life to another.

2012 MARKS THE sesquicentennial anniversary of the legislation that rural historian Doug Hurt called "one of the most important laws in American history"—the Homestead Act of 1862 that provided a single, divorced, or widowed woman the same right as a man to prove up her claim on 160 acres of public land. That agricultural big bang, coupled with the ongoing search for the Midwest farmer's daughter, motivated my own 400-mile trip to the Beatrice, Nebraska, land tilled by America's first homesteader Daniel Freeman and his wife Agnes Freeman, née Agnes Suitor, herself a Midwest rural girl who came of age on the Illinois border. The site of the Freeman claim is now the Homestead National Monument.

It's there that I meet up with ranger Susan Cook, a living legacy of the landmark legislation. Cook grew up on a timber claim homestead in western Nebraska in the 1970s, on the edge of the tiny town of Tryon, where she went to sleep each night in a sod room and received her elementary school education in a one-room schoolhouse with just six other children in her grade. Cook, whose father had been a Midwest farmer and whose grandmother died giving birth to Cook's mother, found herself living what most Americans would consider a frontier existence, sod house and all, in the Carter administration. Indeed, she claims, the very idea that the Homestead Act remained on the books until 1976 in the lower 48, and until 1986 in Alaska, comes as a shock to the visitors she daily educates at the Homestead National Monument.

On her latter-day homesteading childhood, Cook looks back in wonder. "It was so free. My parents were never worried about what was going on, and we roamed the hills close to town. . . . Everybody knew who you were, so if you were getting in trouble, your parents would know right away. Or, if they were tired of you being around, they would send you on your way. I remember we'd saddle horses and just roam."

Cook recalls the sublimely tactile pleasures of the final years of a girl-hood spent on a last frontier, learning how to recognize at a glance the treacherous enchantment of the quicksands along the Dismal River. "We had a group of eight girls who were elementary age and middle school age staying at a ranch house fifteen to twenty miles from everybody else by ourselves, half a mile from the river," Cook recollects. "We got into it [quicksand] and learned how to save ourselves. I'll never forget it, because as long as you relaxed you'd float, but if you moved around, it just pulled you down. If you'd fight, it would pull you. . . . We'd stick our hands in it, and it's really cool because it's just clear water running over sand bottom."

At school the young farmer's daughter marveled at how a farm-wife schoolteacher who rode two miles into town on horseback to teach her classes could somehow command the attention of a room full of forty students of all ages. "There was one kid who had learning disabilities. . . . And one teacher in particular was amazing. . . . She brought in a sandbox and that's what he learned to spell in. Because it was tactile, he got it, and she had tried many different things trying to figure out what would trigger him to learn. She went above and beyond, and that's what I saw of the teachers that we had around us." At other times the rambles that Cook and her fellow country girls indulged in became an impromptu lesson. She remembers, "Two of us girls found a cat skeleton once. We carefully put it in a box and brought it in, and the teacher just watched us try to put it together. We messed with it for several days, and we couldn't figure out why it wouldn't go together. We had all the bones, but we didn't understand the tendons and the muscle. So she let us mess with it for a while, and she finally sat down and explained all the structure of the body, and then helped us figure out how to create a cat skeleton. That was in elementary school."

Lately Cook has been researching the games country kids played in the homestead era; many of them, it turns out, are the same she enjoyed playing with her rural girlfriends in western Nebraska one hundred and twenty years later. "Everything was about everybody winning, about inclusion. So tag was not tag-you-are-out-go-sit-and-wait. Everybody ran hand-in-hand, and then you tagged a person and then they became part of your chain. The goal was to get everybody on your team. So it was all about everybody winning. Red Rover was another big game. If you broke through, then you took people to your team."

Cook feels the games played in contemporary high schools, even those in midsized midwestern towns like Beatrice, have changed, and not for the better where inculcating time-tested agrarian values are concerned. "What we've been teaching our kids is that you have to be the winner, and there's always losers. And what ends up happening is you have kids who aren't as good sitting out all the time. So they never have a chance of getting better because they don't play, and it's about me and it's, 'I'm going to win.' And then they never learn how to lose gracefully. They don't know how to help each other."

In her role as a ranger, Cook speaks to public schoolchildren from across the Midwest, and she reports that, despite their academic precociousness, they're generally abysmal failures at the cooperation-based games of pioneer children. "I make them do chores holding hands, and they can't let go, which means they have to talk to each other on how they're going to do the job. The public schools can't teach that. They're just too competitive," she insists. The lessons of growing up country require some time and care to understand, Cook believes. A case in point is her own childhood sod home, whose yard, she remembers, consisted of such a checkerboard of trees that she could stand anywhere, reach out, and find her hands lost in leaves. She tells me, "As I child I kept thinking, 'What idiot planted all these trees in this yard?' Later she learned of the Timber Culture Act of 1873, which required that the claimant prove trees had been planted and cultivated and that not less than 675 had survived per acre.

For Cook and other latter-day Midwest farm girls, the mysteries of the homesteader's world they inherited proved sufficient to last a lifetime, and compelling enough to create an almost instant nostalgia when, inevitably, they followed the well-traveled trail to Elsewhere.

DR. SARA FAIVRE-DAVIS is one half of what may be one of the most influential Midwest farm daughter pairings in the country. In addition to being the former president of a cloning company and the co-founder of Wild Type Ranch, Faivre-Davis is most recently one of five Barack Obama appointees to the board of directors of the Federal Agricultural Mortgage Corporation, better known as "Farmer Mac." Her sister, Anita Zurbrugg, is the Midwest Director of American Farmland Trust and the head of the Center for Agriculture in the Environment at Northern Illinois University in DeKalb.

Though her appointment by President Obama and likely forthcoming work on the USDA's National Agricultural Biotechnology Advisory Council have her full attention, Faivre-Davis still finds time to take her children with her to the farmer's market where she sells her pasture-fed beef. She's not just a farmer, farm wife, and a farmer's daughter, but a farm mother of two preteen boys, Alex and Eric, a richness she writes about in a blog about raising kids, cattle, and consciousness. In raising her boys on a truly special place—the 333-acre Wild Type Ranch a May 2010 issue of *Edible Austin* made to sound like a veritable paradise of "ethereal, electric green"—she hopes to provide her next-of-kin an approximation of the childhood she enjoyed as a farm daughter.

"Midwest's farmer's daughter was probably my primary identity," Faivre-Davis confesses when I call to ask about the ranch she purchased with the intention that it become a shelter for abused women and girls. "My mom was an avid subscriber to the *Farm Wife News* and regularly sent in poems and pictures. So we all grew up with slogans like 'I'm a farmer's daughter,' and 'Dairy wives have nice calves.' I was lucky and unusual in that I grew up on a corporate family farm. It was a large scale, and I was the last of seven."

After graduating from DeKalb High School in 1982, Faivre-Davis went on to obtain a PhD in agriculture and genetics that led in turn to a series of academic posts, including those at the University of Illinois and at Texas A&M, where she was one of the first molecular geneticists trained in animal science in the country. In each of these academic stints, she bought land, figuring that she was "going to be there forever"—20 acres at the University of Illinois, 20 acres at the University of Nebraska, and 40 acres during her time at Iowa State University. Later, as her first marriage began to crumble in an environment she considered emotionally abusive, she drew strength from the self-reliance she learned as a child. "The farm literally saved my life," she tells me. With her then-husband frequently gone on travels for the genetic start-up company they had formed together, Faivre-Davis was left to put out the hay and take care of the animals who were calving, in addition to providing care to her two kids and looking after a thriving business from home. "Once you have that [independence], no one can take all of your self-esteem away," she maintains. "I lost the ranch in the divorce, but what couldn't be taken away was the self-esteem. That's why the original purpose of this ranch was to give that feeling to other girls and women." Buying her own 333-acre ranch as a safe haven after the divorce felt right to Faivre-Davis, in part because she was returning to a recipe she knew back on an Illinois farm that at one time totaled nearly 7,000 corn and soybean acres. "I saw my whole family working on the family business," she recalls of her childhood. "If I took a nap, I'd know my mom would be out in the garden."

Her own broken marriage sharpened Faivre-Davis's recollection of her wonder years, and the heritage she wanted for her children. "They needed to be grounded," she recalls of her decision to buy the pricey ranch she hoped would prove restorative. "They needed to understand that they can't have everything they want. Before I was divorced, both of us had six-figure salaries, and we lived on a ranch, but nobody had to work. . . . When the company went public and after my kids were grown, I was going to establish a ranch that would be a long-term residential facility for women and kids coming out of domestic violence situations. Because there's nothing like working with cattle, working with animals, driving a tractor, producing your own food, and doing all these physical things to give you a sense no one can take away."

While the dream of a farm tailor-made for a girl's recovery never quite came to pass, Faivre-Davis takes heart that her farm still serves as a rejuvenative place for the many nieces and nephews who yearly bunk there, and in that way she believes she and her second husband have "fulfilled the mission in a different way," by sharing the healing power of a rural upbringing with a family full of eager children. After six years on the ranch, she can already see the difference in her own two boys. "They are really connected with their livelihood and their food," she tells me. "Especially because we are in the local food network, and we raise beef and sell it locally. . . . A lot of times we sit down to supper, and aside from things like bananas and spices, we either grew it ourselves or we know who grew it. The boys have no illusions about what goes into putting food on the table."

Eric and Alex work on the ranch, first for chores, and, then, if they wish, for spending money. Faivre-Davis believes the kids are learning something valuable about a balanced work ethic. On Wild Type Ranch mornings are for work, and if the job gets done, afternoons are for play. The boys seem to thrive on the regimen. "Because they live in two different households, it's very obvious to them that that's not the way everyone lives, because they don't live that way with their dad," Faivre-Davis confides. "My oldest son has said to me, 'You know, mom, I think I'd even work beside you if I wasn't getting paid, because there's just something about working that makes you feel good.'"

MY OWN BABY boomer mother shared precious few of her coming-of-age memories with my sister and me as we grew. It hadn't been emotionally easy—that much we understood intuitively—but at the same time it hadn't been hard—at least not in the backbreaking, physical sense. We also knew, somehow, that we were living the same life our mother knew, and in that sense we could be spared the recitation of the many goodnesses of growing up country.

"It seemed safer," my mom explains when, nearly forty years later, I ask her about her decision to move my sister and me back to the farm. "I knew eventually it would be the same story . . . that you guys would one day leave. You were little. It just seemed safe, and we were closer to family, and I was starting to appreciate the country, the woods, the privacy."

Mom hadn't always embraced the virtues of raising a Farmer Jane—far from it. When she hit puberty, town, some two miles distant, seemed a strangely attractive universe to her. As a teen, she grew to feel isolated and envious, but as a girl, the farm life seemed an unending adventure. "An overall view might be that I really liked living on a farm until I was about 11 . . . that's when I stared realizing that the farm to me as a girl was boring and that my friends were having a great time in town," she confides. "And being out on the farm I started to feel really isolated. . . . After about 10 or 11, when I realized there was a big world out there, and I wasn't part of it, then I didn't like living on the farm . . . in fact, I started to hate it. That feeling carried with me for a long time, both being a farmer's daughter and being married to a farmer at a young age."

On my father's side, my aunt Barb, three years my mom's senior, tells a similar story. "Once I started school I always felt that I was missing out on something. Town kids could easily walk to friends' houses, to the park, to the playground, to the drugstore for pop and candy . . . our bus ride home was about forty-five minutes, but seemed like an eternity. Town kids could walk home in a matter of minutes. I always worried about missing the bus and being stranded in town . . . that was terrifying for an extremely shy little girl. When I was in elementary, I also felt that country kids were not as good as town kids. But maybe that was all in my head."

In her head perhaps, but also on paper. Even Barb, perhaps the most accommodating of a genial family of farmer's daughters, sometimes found irksome the role of growing up a girl on the farm. Exhibit A is the sole surviving written artifact of those teenage resentments, a composition my aunt wrote in the early 1960s for her literature teacher, Mrs. Pruess. She titled it "I Hate Being Bossed." It read:

> It can't be escaped. If not at home, I get it at school. My parents are forever yelling at me to do something: anything.
>
> *Sweep your room.*
>
> *It's your turn to do the dishes.*
>
> *Burn the waste paper.*
>
> *Feed the cats.*
>
> *Dust the furniture.*
>
> *Clean the bathroom.*

"Oh, the pain of it all," a teenage Barb complained. Where her parents' demands left off, her teachers' began. The remainder of the essay adds to the litany:

> *Barbara, stop talking.*
>
> *Read the next chapter. It's only 200 pages, so have it done for tomorrow.*
>
> *Get back in line.*
>
> *For tomorrow, do the next 40 pages.*
>
> *If you haven't any work to do, I'll give you some.*
>
> Never mind. I have enough.
>
> When will it ever end?
>
> Must I always be a victim?
>
> *Barbara, get busy.*

As if endorsing the pitiable position of the farmer's daughter, Mrs. Pruess penned across the top of the paper, "Very Clever!," awarding my aunt's lament a 99/100, and punctuating the stellar mark with an oversized "A."

Halfway across the state, Rachel Garst, granddaughter of one of the most famous American farm families, was having a similar experience.

Garst's grandfather, Roswell "Bob" Garst, had made international news in September of 1959 when he hosted Russian Premier Nikita Khrushchev on the family's 2,000-plus acre show farm near Coon Rapids, Iowa. "Iowa farmer Bob Garst Has Much to Show Nikita" the headline of Richard Orr's story for the *Chicago Tribune* read. Orr described Bob, the family patriarch, as a "talkative, shrewd . . . blunt spoken, suspender-wearing man who doesn't hesitate to offer unsolicited advice to strangers." These very traits had been attributed to Khrushchev, Orr claimed, "which could be the reason the two get along." The Garst farm was by no means typical of the average midwestern spread in 1959, Orr carefully pointed out, adding that the "air-conditioned six-bedroom home and nearby swimming pool" attested to the difference.

As a teen Rachel Garst came of age in the shadow of her larger-than-life grandparent in a community, Coon Rapids, where to this day an American flag, a billboard, and a giant mechanized ear of corn announce to visitors that they've reached "The Heart of Corn Country." In one sense the outsized reputation of her famous family meant Garst's upbringing was different than other daughters of the farm, but in another she was one of hundreds of thousands forced to reckon with a farming patriarch whose stature somehow eclipsed the long-lived and equally energetic farm women of the family. "My mother has probably been my most important role model, followed by my oldest sister and my aunts," Garst tells me. "These women have always been active and engaged in the outside world, and politically involved, more so than the men in my family, who seemed more focused just on the farming and business angles. My mother taught me to be a good citizen in the broadest sense of the word: to vote, to pay attention to politics, to care what was happening with the city council and the school, and to volunteer. She also managed a home that included real meals, sitting down together, celebrating holidays, and all of the rituals that are so important to human beings as familial and social creatures. My mom and grandma also had way better manners than my dad and my grandfather; I appreciated that about them. Farm men sometimes can be crude."

"It was difficult to deal with the resentment of certain people who assumed that I led a privileged and coddled life as a 'Garst,'" Rachel recalls. "One way my parents dealt with that same pressure was by training me to work, and to work just as hard as anyone else, be it delivering newspapers as a young child, or later as a teenager being sent out with a shovel on thistle crew, or put on a horse or chute to work cattle. If I signed up to do farm-work, I was expected to be on time, not be a whiner or expect special privileges, but to work to my best ability."

Garst worked for her parents' in-town cattle business during her teenage years, and, for her, childhood stints on the farm had special appeal. "I liked best being outside when the sun came up and also when it went down. I liked watching the wind ripple across the pastures. I liked the excitement of chasing a wild cow and running her through the chute. I liked the toughness of the cowboys (and cowgirls!) that worked for my dad, and the camaraderie of being on a 'crew.' I liked the toughness I felt in myself as I struggled to stay awake; or saddled a horse; threw bales of hay; manned the chute; or dealt with the heat, cold, and wind. I loved everything about it. Working outside as a farmhand or indoor working cattle were the best jobs I ever had. Walking outside is great, but working outdoors, being outdoors with a purpose and a goal that engages you with the natural environment, is the most amazing experience ever."

While Garst has since returned to her native Midwest, she was eager to leave Coon Rapids, taking flight before she completed high school for a series of study abroad experiences that served as a seedbed for an adult life lived beyond US borders. "I enjoyed the anonymity of living far away from Iowa, where my last name is well known and sometimes causes people to treat me with resentment," she recollects. "I really enjoyed being far enough away to rise or fall on my own merits, just as me, removed from any family pressures or reputation."

For my mother the pleasures of a farm daughter girlhood were poignant precisely because they were self-directed. In her world the only real "family time" was likely to come in the late evening around the dinner table, especially in the summer, so she learned to make her own fun and to depend on her siblings to help find it. "We just used to find stuff to do. . . . We used to play in the hayloft all time. We'd climb to the top of the ladder and swing from the ropes. We could have so easily gotten hurt. But it's almost like we had this innate savvy. We would go down to the creek and be gone for hours—we were little, like 6, 8, 9, and we would wade in the creek for miles. We never got hurt. We got leeches on us, but we always made it back."

In the days before she heard the siren songs of jobs, dates, and afternoons at the swimming pool, my mother would take fun where she could get it. "My brother had a go-kart," she remembers. "Of course we had no computers, nothing like that . . . so we liked to kind of stay out from under our mom's feet, so she didn't find jobs for us. We hung out in the barn. . . . Our big thing in the spring was that we knew the cats had kittens, so we would follow them, and we would spy, and we would hide behind things, and it would take weeks to find the litter. Then we'd find them, and she'd move the litter. Other times, we'd take branches and make stick horses. And we'd just gallop around, and we'd go across the road and make trails for our horses to jump. In the evenings," she continues, "we always went out and played games outside: hide and seek, catch the fireflies."

Like Garst, my mom traces many of her most poignant farm memories to her mother, who assumed the responsibility, however imperfectly, for making memories. "I look back on it now," my mom tells me, "and the Easter Bunny and the Tooth Fairy and Santa Claus were all women! But this is something my dad did that was very nice and thoughtful. He bought us a pony, and he walked five miles to get the horse, and then he walked back five miles with it. It was our Christmas surprise. And so as he was coming down the road, mom was watching, and all of a sudden she had to shut the curtains. He walked this horse all the way home, and put it in the barn, and then on Christmas morning we went out and there it was. As farm kids, to have our own pony, even to share among five kids, was just overwhelming joy. . . . It turned out to be a bucking and biting and kicking horse, but he was *our* horse."

For Garst, as for my mother, the impetus and willingness to return to the country life began with her children. "Foremost in my decision to return to Iowa, and particularly to rural Iowa, was my ongoing emotional connection to the landscape of my childhood, and my desire to give my children a farm upbringing," Garst says. She hoped her kids would be called to work outdoors and to have the kind of formative experiences she had, which she now regards as essential to moral, spiritual, and physical development. "I am extremely worried that 99 percent of current young people are cut off from that opportunity," she tells me. "For that I blame our changing agriculture, where fewer and fewer famers are needed, but also our car culture, television, computers. . . . I see families in Coon Rapids that pick up their kids from school rather than let them walk even four or five blocks. That just seems crazy to me. . . . I also see a safety psychosis affecting the entire country, in which children are not even allowed to play in the neighborhood park because of overblown fears of 'sex predators.' For me the benefits of engaging with the outdoors far, far outweigh any dangers that may be encountered from nature or from other people. I wish to God every kid in the United States could be raised on a farm, regularly visit a farm, or at least have a garden."

THE IDEA THAT a rural upbringing could, if properly carried out, approach the ideal was widely asserted in the years immediately before and after the Great War, even as the health of rural civilization was called increasingly into question. Distilling the sentiments gleaned from rural mothers who had written her from across the country during the Roaring Twenties, author-scholar Mary Meek Atkeson summarized, "She [the farm woman] feels that nowhere in the world can her children learn those moral values which are always the highest good in life, and develop real stability of character, so well as in the country home. They can learn the secrets of life, and responsibility in caring for helpless young animals, and can realize the suffering brought about by the neglect of an obligation. And this will secure their success and happiness in after years, wherever they may be."

While Atkeson's study focused on the farm mother, Martha Foote Crow's *The American Country Girl*, offered space for the country's turn-of-the-century rural daughters to speak for themselves of the pleasures and pains of the so-called dirty life. Crow devoted an entire chapter to Illinois farm girl Elizabeth Wilson's poetic accounting of the unique pleasures of her American rural girlhood. Wilson's invocation read:

> I love the taste of thorn apples and sweet acorns and sumac and choke-cherries and all the wild things we used to find on the road to school.

And I love the feel of pussy willows and the inside of chestnut burrs.

I love to walk on a country road where only a few double teams have left a strip of turf in the middle of the track.

And I love the creaking of the sleigh runners and the snapping of nail-heads in the clapboards on a bitter cold January night.

In the first cool nights I love the sound of the first hard rainfall on the roof of the gable room.

And I love the smell of the dead leaves in the woods in the fall.

I love the odor of those red apples that grew on the trees that died be-fore I went back to grandpa's again.

I love the fragrance of the first pink and blue hepaticas, which have hardly any scent at all.

I love the smell of the big summer raindrops on the dusty dry steps of the schoolhouse.

I love the breath of the great corn fields when you ride past them on an August evening in the dark.

And I love to see the wind blowing over tall grass.

I love the yellow afternoon light that turns all the trees and shrubs to gold.

I love to see the shadow of a cloud moving over the valley, especially where the different fields have different colors like a great checker-board.

I love the little ford over Turtle Creek where they didn't build the bridge after the freshet.

I love the sunset on the hill in Winnebago County, where I used to sit and pray about my mental arithmetic lesson the spring I taught school!

Still another farm girl epistle of the time described the advantages of a farm upbringing thusly:

The quietness and peace which permit of one's greater spiritual and mental growth, the abundance of life, plant and animal, which chal-lenges the mind to discover its secrets; the rocks and streams which call out to one for study and discovery, the beauties of the sunrise, the clouds, the sunset, the moonlight, and the far off stars—these call to our spirits to penetrate their mystery and lift up our souls to those levels above the commonplace where we commune with the Maker; the hills and the wide expanses make us reverent and teach us to walk humbly and patiently; the clean sweet air gives us health and strength of body and soul; and the freedom from restraint by formalities and conventionalities permits the development of the person in a sane and natural way.

Another turn-of-the-century soil sister captured the metaphysical inheritance of the farm, insisting, "Farming is creative. . . . On the farm both body and mind are exercised, therefore both are kept nearer a normal level. We have fresher, purer food and air; freedom from foolish forms and ceremony. We are nearer to God."

Another of Crow's rhapsodic farm girls enthused:

The attractiveness of farm life lies in as many, diverse, and wonderful things as the breadth of the individual girl's mind can comprehend and enjoy. To some the sense of freedom in country life is a large means of happiness. The feeling of exultation in the far sweep of vision, the glorious sunsets, and the movements of the clouds in the wind and the coming storm. Then there is the pleasure in seeing and helping things grow, in the frolic of the lambs in the spring, of the colts at play, and in the young plants sprouting and growing in the summer showers and sunshine; especially if you have pulled the weeds and hoed about them yourself. Frequent outings to the lake or river for an afternoon or evening holiday with bathing and canoeing in the afternoon and a bonfire in the evening with a group of friends to toast marshmallows or roast corn, and later with stories and songs, add much to the pleasure of farm life. Then there is the quiet and peace of the country where one may be alone at times and think. In the country there is a more compact home life than anywhere else, for each member of the family is working together for the home.

Last to sing the praises of the agrarian lifestyle was a 19-year-old daughter born to a 160-acre hill farm, who wrote, "Farm life to me is attractive . . . because on the farm one has freedom that cannot be gained anywhere else in the world. One learns the habits of birds and animals and one comes in touch with nature and hence with the Creator himself." While the writer conceded that "there may be reasons why one might desire to leave the farm," including hard work and little pay, she insisted "anyone leaving the farm will never be happy while away, and will soon learn that there is no place in life like the farm."

IN THE NEW millennium memoirs of farm girls who left the barnyard behind vastly outnumbered accounts written by those who had stayed. More popular still, given the dramatic demographic shift away from girlhoods spent on working farms, were participatory accounts written by urban folks who decided to the give the farm a try, such as Kristin Kimball's *The Dirty Life*. Meanwhile, almost completely missing in action were accounts of farm daughters who had grown up on working farms in the 1970s and 1980s, girls who, as a consequence of their own demographic exceptionalism, found themselves without the audience guaranteed baby boomer authoresses who had pitched hay in the era of Jack and Bobby Kennedy.

I begin my search for the thoughts and motives of farm daughters 40 years old and younger with my sister, asking her to recall the most lasting effects of growing up on a Midwest working farm. Natasha Jack-Hanlin grew up on a 500-acre Century Farm in Cedar County, Iowa, and, like many Midwest farm girls, went on to work in "helping professions," first as an elementary schoolteacher and later as a human resources manager. Nearly a thousand miles removed from her home ground, she finds herself processing more thoughtfully the lessons and memories of a farm girlhood as she raises her two boys:

> Some of my favorite memories are visceral, sensory memories. The smell of coffee and cigarettes on my dad. Laying in a wagon full cool, smooth corn. Lightening bugs flashing on and off on a still, summer evening. The sound of a rustling cornfield. These brief, sensory images remain very important to me. I think it is because, as an adult, I now see all the holes that existed in our family life, both immediate and extended, and the pain and hardship that was there, but the sen-

sory experiences can't be reasoned away. There is no badness there, they just are. Many of my favorite memories of my father are centered around things we did together around the farm—combine rides in the fall, 'working' with him in the shop, trips to the implement dealer, having him teach me how to make tables for my Barbies out of four nails and a piece of 1 x 4. Because he was always so busy, these brief moments were really important. My least favorite memory would be missing him so much, as he didn't have a lot of time to spend with us.

SARAH RYERSON-MEYER'S grandmother and grandfather, Glen and Doris, moved onto the homestead on which her great-grandparents had fashioned a home eight miles outside of Northwood, Iowa. As a girl Sarah enjoyed 80 acres around which to play leading lady on a set of a half dozen thoroughly romantic outbuildings, including a chicken coop, a well house, and a defunct washhouse where her ancestors had scrubbed their laundry. Sarah's grandparents stayed on the farm late into their lives, long after a plan for her parents to buy the family ground fell through. Eventually, her brother bought the house and 10 of the acres while her aunt inherited another 70 acres. Sarah's warm recollections are wedded to the evocative setting and narrative the farm planted in her psyche:

> There was a lot of quiet on the farm. Being outside was one of the best parts of the visit, and it wasn't a highly traveled road—at the time it was still gravel. It was also quiet in the house. My grandparents were very loving people, but they were not people of many words. I remember many hours sitting with them in the living room, just sitting and feeling very comfortable and very loved—with no words.

> I used to climb in the hayloft and marvel at the huge-bodied spiders. Grandpa would feed the cats in the barn the same way every night after supper. He gathered the food scraps from a pail Grandma kept in the "breezeway"; he would take scraps to the wash house, where he added dry food from a big bag, close the door behind him tightly, and

walk to the barn calling, "Here kitty, kitty, kitty, kitty." And the cats would line up behind him. He did not act like he loved them in the least, but he always fed them.

As a child I viewed the farm as a somewhat dangerous and also very exciting place to spend time. It was the wildest place I could get to. There were patches of trees that were far from the view of the house where I could climb and wander on my own. There were animals that I had to be aware of and respectful of. My grandpa kept a big Black Angus bull, and I distinctly remember one day when I came face-to-face with the bull unexpectedly. My mom was having coffee in the house, and I walked out to the barn, evidently looking at the grass most of the way. When I looked up, the bull was in the middle of the yard, outside the fence. I had been told many times not to run or make any sudden movements around the cattle, and somehow I just kept walking and made a U-turn right back to the house. What really got my attention was that my mom, my grandma, and my grandpa were waiting for me at the door. They had seen the bull too late and had been afraid to yell a warning at me.

Because of time spent on the farm I knew what a rendering truck was. It didn't happen often, but when Grandpa had a sheep die, it would lay in the yard near the driveway waiting for the truck, legs stiffened at an odd angle. It was something that was difficult to look at, and yet I was drawn to look at it—maybe my first very close look at death.

SARAH UTHOFF'S FAMILY goes back six generations in rural Johnson County, Iowa, deep enough that it makes her long commute to work as a community college reference library in nearby Cedar Rapids well worth it. She's spent a lot of time thinking about gender on the farm, and about the importance of active rural female role models. She says she loves living on a mixed agriculture midwestern farm where she and her family raise Hereford cattle, Suffolk sheep, and chickens, and where, as a child, she had a steady stream of barn cats, wildlife, and birdlife to keep her occupied. "You are never allowed to be bored," Sarah says. "Busy and exhausted—yes. Bored—no." She remembers:

> I know there are some people who like to live in town, but it's not for me. They say it takes a village to raise a child; it takes a whole family to run a farm. You spend a lot of time with your family, and not only with them, but working together on a project. You not only get to know them better, you get to know them as part of a team that you are an important member of. My other set of grandparents lived in town, and when I was little I used to go to stay with them sometimes for a couple of days. You feel so cooped up and boxed in town. You have to be careful how loud you are and always be aware of people around all the time. I hated that feeling.
>
> I did a lot of things that town girls wouldn't. I showed Herefords at the county fair, so I had to take care of them and break them to lead (my brother always tells people, "Don't make her angry; she can break her own calves"—which seems to impress them far more than it really should). I helped take care of all the animals. I helped make hay. I helped with the garden. Whatever the current project was I just naturally came along and helped.
>
> There is a difference in attitude too about a lot of other things. For example, town girls seem a lot of the time to need to be entertained, but for farm girls there is always something to do or something to be interested in. That's just a start. This very weekend I had an exchange on Facebook commenting on a mutual friend's post. This person apparently got her information about animals from Disney cartoons and was saying how opossums are our friends. Opossums eat chickens and eggs; people want to keep their chickens and eggs: how is that a basis for friendship? I've dealt with a lot of opossums, and trust me, you don't want to be taking a feel-good attitude with them. Town people in general a lot of times seem not to understand that animals aren't stuffed toys they can play with.

AUDRA BROWN IS one of many generation X midwestern farm girls who traveled widely before putting down roots in environs very different from those in which she grew up. Audra's parents fell into farming as a second-choice profession, her mom forever yearning to be a nurse, and her parents discouraged her and her sister from agriculture, citing the lack of return on investment, though she insists that "they wouldn't have stopped us if that's what we wanted to do." Audra is the first to admit that, like a number of farm girls who came of age in the 1980s, she wasn't tasked with regular chores, though she did tend 4-H pigs and looked after the family's fifty barn cats. She shadowed her father as much as she could, and often found herself wondering whether she would have been asked to contribute more had she not been a girl. She reports that for years she "resented her gender" as a result of those nagging doubts. Her playtime consisted of "taking care of invisible animals and riding imaginary tractors." A special education teacher now living near her husband's air force base in Charleston, South Carolina, Brown's recollections of her Midwest home orbit around its spiritual resonance and the sense of longing she feels for it. She recalls:

> I grew up on a one-hundred-acre farm in Louisa County, Iowa. It was paradise. I always cherished our farm because many farmhouses seemed plunked down in the middle of a cornfield, but our house was nestled between several acres of timberland at the end of a gravel road. Our closest neighbor was a mile away. Long creek . . . ran along at the edge of our property and for an adventurous tomboy like me, there was no end to the fun. I roamed every hill and dale with my two dogs collecting bushel buckets of unique treasures: arrowheads, animal bones, hedge balls, tadpoles, and rocks. During the summer months I would pack a lunch, head to the creek, and come home around suppertime.

> I think we [farmers' daughters] have a deeper spirituality because we are surrounded by God's handiwork. When I was a girl I would sit under a particularly tall oak tree and take time to appreciate the natural world. I considered myself lucky to be able to enjoy the animals, the plants,

the seasons. Every living thing had its purpose and its place, and I was part of that great plan. I could sense God all around me, and I had such difficulty understanding how anyone could deny His existence—the evidence of a master designer was overwhelming. As an adult, I've found a Bible verse that sums up those feelings I had as a young girl sitting under that grand old oak: "They know the truth about God because he has made it obvious to them. For ever since the world was created, people have seen the earth and sky. Through everything God made, they can clearly see his invisible qualities—his eternal power and divine nature. So they have no excuse for not knowing God."

I did resent being a girl for a very long time. I wanted to do the chores that my dad was doing. Even in the 1980s most girls were expected to work indoors cooking, cleaning, and baking. Although I did enjoy doing those things, I really wanted to be outdoors working the farmland and tending to the animals. In the back of my mind I felt as though girls were not capable of doing this sort of work.

Whether or not every farm girl was in 4-H (but most everyone was), we were raised to help other people. Years ago that's how people made it while living on the land—they depended on one another—and that thinking has continued from generation to generation. This thinking has served me well, particularly in my work. Work on the farm was never a nine-to-five job, and even though I didn't actually do a lot of farmwork myself, I saw my parents rise before the sun and continue to work well into the night. Throughout my job history I have been awarded all kinds of "best employee" awards, and I am usually promoted faster than others. I don't think it is because I am smarter than anyone else—it's due to my inherited work ethic. I tend to show up to work early and stay late without thinking about compensation. Farmers work until the job is done and I do, too.

The downside of this is I have been taken advantage of—a lot. I grew up trusting others. We were such a small community that we were like extended family (and actually there was so much intermarrying that everyone was related somewhere along the line). I didn't see much crime or violence. People didn't lie, cheat, or steal. When I went away to college, I gave a lot of money to those in need, and I was often swindled by people who weren't telling me the truth. I have smartened up through the years, but I still give to people. My husband has chastised me many times for picking up people on the side of the road and taking them to where they need to go. I have often emptied my pockets or given away my groceries. . . . If this is a bad thing, so be it.

"**I WONDER SOMETIMES,**" writes author Kristin Kimball, "how Jane will regard her childhood. I am aware it's not your average one, at least not here and now. We spent her second birthday, for example, butchering rabbits. . . . When I meet adults who grew up on farms, I quiz them on their upbringing. The answer is never lukewarm. It's either painted in golden colors—the perfect way to grow up—or described as pure drudgery, no childhood at all. The split seems to run fifty-fifty. I love this farm and the life that comes with it. I love that it makes me feel rich even though we're not. I love the work. I figure the best we can do is share that love with Jane, and hope she loves it, too."

The recent bumper crop of farm memoirs demonstrate that new generations do find their way onto the farm, discovering for the first time the joys of raising Jane there. And when they do, they join an ever-dwindling but highly purposeful minority trusting the nurture of their children to a timeless reciprocity between the land and those who love it as wholly as a child.

Celeste Pille, the artist of "Farm Life," is a Midwest farmer's granddaughter and graphic artist living in greater Omaha, Nebraska. She recalls of the sale of her family land: "It was the early to mid '90s when my grandparents sold the farm. It wasn't a sad occasion, but seen through my eyes at the time, you would think it was. I was gloomy that day, and I think a lot of it had to do with the fact that I was just a kid. Your world is so small . . . any change seems like a global event. I completely associated my grandparents and our visits with the farm, and to me the auction day was The End. Really, though, Grandpa still had land and rented it out, but the land with the house where my mom grew up and all the machines went that day."

Illustrations by Celeste Pille.

My grandparents sold the farm. They held an auction and I spent the day scouring the fields for treasures I could afford. I found a few items of interest and ended my search by weaving a crude chain of clover flowers. I had an old, glass doorknob that led nowhere, a button that secured nothing, and a necklace that wouldn't last.

They moved. Rows of crops became rows of houses and the smell of earth was replaced with the smell of mowed lawns.

I would miss the farm-fresh corn. Canned corn should be considered an oxymoron.

The next time we visited Grandma and Grandpa I knew there would be no trip out to the farm. There would be no ride on a lumbering tractor and cooing over animals. I resigned myself to a farmless existence, so when Grandma placed a hot plate of corn on the cob on the dinner table I was very surprised. I don't know what best mirrored my joy—the flourish of the steam or the sumptuous yellow.

CHAPTER FIVE

THE CHORES OF
BEING A FARM GIRL

MIDWEST FARMER'S DAUGHTER Sarah Evans's song of a working farmer's daughter itching to kick the dust off her boots, "Suds in the Bucket," went Gold in 2005 on the RCA label. Five years later the song's stunning entry into the country cannon was cemented by its selection for the cult video game Rock Band, allowing girls across the country to indulge in a simulated sing-along of a head-over-heels farmer's daughter who left home in "the blink of an eye," leaving suds in the bucket and clothes drying on the line. In an instant Evans's catchy tune of leaving chores undone and the farm life unlived had become a classic, earning its place in the Country Track Pack alongside Willie Nelson's "On the Road Again," Kenny Roger's "The Gambler" and Kenny Chesney's "She Thinks My Tractor's Sexy."

The song's lyric hook, and the crux of its emotional appeal, concerned an 18-year-old girl stuck with the farm's inglorious work. The music video opens showing Evans astraddle an old-fashioned washtub, before quickly dissolving into the image of a high school-age country girl—presumably meant to represent a younger version of the singer—wearing braces and a white gingham dress, as a blue jeaned boy pulls up in a white pickup truck. In the blink of an eye viewers see the washboard dropped in the bucket and the image of a barefoot girl sprinting to the chariot of the down-home Prince Charming set to take her away. The video next cuts to the image of the bewildered parents scratching their heads in the kitchen, wondering how an 18-year-old-girl could, as the song puts it, just up and walk away.

A hundred years before Evans's chart-topping single and subsequent tabloid divorce from her real-life husband that MSNBC called "sordid," the debate about the rightness of a farm girl's upbringing likewise centered on a country girl's labors. In the mid- to late 1890s, as devastating weather and deepening recession hastened rural-to-urban migration in the Midwest and Great Plains, rural America wrestled mightily with the question of what it could fairly expect of its daughters. "Although many children willingly helped their parents on the farm without damage to their health," rural historian Doug Hurt reports in *American Agriculture: A Brief History*, "precious childhood years could easily be lost in drudgery."

Anti-child labor resolutions had been drafted in the industrial East since the early 1830s, when the New England Association of Farmers, Mechanics, and Other Workingmen resolved that "children should not be allowed to labor in the factories from morning till night, without any time for healthy recreation and mental culture." Most of the era's legislation focused its regulatory power on city-dwelling employees who had secured "gainful employment," a term that functionally excluded the kind of farm labor generations of rural parents had come to regard as a natural expectation of their children. That many of the organizations then calling for tighter regulation of child labor, such as state Consumers' Leagues and Working Women's Societies, were run by urbanites who had long since abandoned the farm further exacerbated the issue in the rural press.

In October of 1898 the *Iowa Homestead* printed pages of deeply felt letters from its readership concerning what it called the "financial position" of the farmer's daughter. "The condition of the farmer's daughter now is that of a slave," wrote Herman Jahnke of Regina, Wisconsin, "and will remain so until the farmer gets his eyes open." Jahnke insisted that a girl be given at least 50 cents an hour for a maximum nine-hour work day and be allowed at least one year of high school. Farmers' daughters themselves, including Katie G. Williams of Manhattan, Kansas, put pen to paper to demand an allowance for their and their mothers' labors. Williams pointed out, "The husband and brothers are likely both to say, 'Oh, you don't need the money; you don't know the value of a dollar.' If they don't know, and are never allowed to handle money, how can they learn?" Flora Wood of Waubeek, Iowa, agreed, though the reasons for her support for wages for the farm girl were more practical. "If a farmer owns his farm, and it is large enough to keep the boys at home, it surely ought to keep the girls. Mother needs the girls, and no real, dutiful daughter would refuse the gift of a nice cow, or a calf, even to keep and sell for her own." The time had indeed passed, Wood went on to note, when rural girls and women would "submit to injustice." They had learned, she said, to "demand that they have their rights, and an occasional dollar, without asking for it meekly."

The dissenters in the *Homestead* were few. A farm daughter calling herself "Red Willow Country Girl" dispatched her opinion from McCook, Nebraska, opening her letter, "The lot of a farmer's daughter it seems to me is not so very deplorable." Ever since they were "tiny tots," Willow Girl avowed, she and her farm daughter sister had saved their own earnings. Based on sales of rags, turkeys, and ducks, they had, with the help of their father, been able to afford a year at a teaching college. "This year finds us both with good certificates and good positions. We are still farmer girls, however," Willow Girl noted, "and as we are the only girls in the family, are assistant housekeepers as well. We try to arrange our work so that one of us is at home with mother most of the time." With what they termed their "carefully hoarded treasures" the sisters were planning a still higher education.

Only the letter received from C. M. Mather of Clarksville, Iowa, dared to resist the sentiment of the hour, arguing "the law recognizes the child, male or female, as belonging to the parent wholly and solely. The parent is obliged to feed and clothe the children and is entitled to all the benefits of their labor." A farm daughter's worth was relative, Mather maintained, to her character. "If she is one of the shallow sort that can't think of anything else but gadding, attending dances, and parties, discussing this fellow and that fellow and having a beau at sixteen or less and is conversant with all the slang of the day, while her mother works her fingers to the quick and her father's money is wasted to make her shine, then she has no financial position, and I hope she never will."

Sentiments like these, considered harsh and unforgiving by many urban Americans, prompted, in large part, Teddy Roosevelt's Country Life Commission of 1908. Farm women found a powerful ally in the man Roosevelt appointed chair, a Midwest farmer's son named Liberty Hyde Bailey, whose sympathies for rural women had been based on a careful gathering of feedback under the auspices of Cornell University's Farmer's Reading Course. In a letter dated 1901 and addressed "To the Farmer's Wife," Bailey declares, "In all the vocations of life, there are none in which success depends so much on the wife as in farming."

Bailey urged farm women to "talk back" so that he and his colleagues at Cornell might get a handle on the most vexing issues facing voiceless farm women. Return postage would be paid by the state on behalf of the College of Agriculture, he assured his thrift-minded audience, if only they would send in their feedback. Elsewhere in his agricultural extension outreach, Bailey mobilized women's efforts for the betterment of rural life in general. On the subject of sanitation and running water in the country, for instance, he played the role of provocateur, advising country girls to decline all proposals from suitors who failed to promise running water and a usable kitchen once married. In section six of the *Report of the Country Life Commission*, entitled "Women's Work on the Farm," Bailey lodged emphatic claims on their behalf, asserting the "burden of . . . hardships falls more heavily on the farmer's wife than on the farmer himself." The relief to farm women, wrote Bailey in an oft-quoted statement of the time, "must come through a general elevation of country living. The women must have more help. In particular, these matters may be mentioned: development of a cooperative spirit in the home, simplification of the diet in many cases, the building of convenient and sanitary houses, providing running water in the house and also more mechanical help, good and convenient gardens, a less exclusive ideal of money-getting on the part of the farmer, providing better means of communication, as telephones, roads, and reading circles, and developing of women's organizations."

In part the strategy of the Commission was to sway a generation of intransigent rural men by shaming them into remediative action. In his 1913 column for the *New Castle (PA) News* entitled "The Queen of the Farm and Her Work," farm photographer H. Winslow Fegley followed Bailey's lead in drawing a stark dichotomy aimed at the era's menfolk. Leveraging the farmer's daughter, he wrote:

> On many farms it is still customary for the housewife to split the wood she needs in her kitchen stove. In such homes nobody needs to ask the question why the daughter left for the city to work at a loom in a silk mill for $7 or $8 per week. Her mother's wrinkles and her growly expression can answer this question. On the other hand, the farmer who sees that his coal bins are full, his woodpile always replenished, the wood chest filled every night for the next day's necessities, and who provides a gas stove to iron the clothes in summer days, when the ordinary wood fire would make a furnace out of the farmhouse kitchen is the identical who can enforce the new slogan, "Stay on the farm," without radical methods.

Galvanized by the Country Life Commission's report, academics and journalists fanned out across the countryside in the 1910s to determine what might curb what Martha Foote Crow, dean of women at Northwestern University, called "the funeral procession of the nation cityward." With renewed attention paid to farm girls and women, scholars expanded their right-living and child-rearing treatises to include the fairer sex. Kansas State Agricultural College philosophy professor William A. McKeever mined the theme for several books, most notably 1912's *Farm Boys and Girls*, wherein he offered advice to parents on how to create a "splendid crop of rural boys and girls." Published as part of Liberty Hyde Bailey's Macmillan's Soil and Science series that included such titles as *The Forcing Book*, *The Pruning Book*, and *The Training and Breeding of Horses*, the monograph's agrarian context was explicit. Without a whiff of irony McKeever titled one section "Do you own your daughter?" having in mind the farmer father who "seem[ed] to regard his 23-year-old daughter as a chattel." The "hurry and work and the isolation of the ordinary country home," McKeever argued, tended to "foster an over-serious disposition" in girls whose laughter, once their hallmark, had become "too infrequent." By analogy, insisted the professor, the farm girl was the "tender plant of the household" who could, he acknowledged, be "kept alive on work and study alone, but for beautiful and symmetrical growth other elements of character nourishment" were indicated. For a list of desired traits McKeever referred his readers to his book's first chapter, wherein he listed "usefulness," "social efficiency," and "religious interest" among the farm girl's prescribed virtues.

Elsewhere the monograph's headings proved equally revealing of the issues then occupying rural minds regarding the "girl crop," including an overarching question where girls were concerned: "Should there be an actual investment?" Number one on the list of ways to keep the girl on the farm, wrote McKeever, was "teach the girl to work." Only hands-on toil, he maintained, would allow the farm girl to one day direct others in manual labor; "the strength of our democracy is much dependent upon the character of our women," he declared, adding that the trend then developing toward a "leisure class" of girls and women would be as menacing to future "social solidarity" as the older tradition of "keeping women in ignorance and servitude."

Due in large part to the efforts of the County Life Commission, within a generation it became an almost patriotic duty to criticize traditional commodity farmers for mistreating their young women. Farmers could be convinced, the Commission realized, to advocate for rural daughters if liberating those daughters meant farm men having to choose between what they considered the lesser of two evils: greater gender equality, on the one hand, and wholesale abandonment of the farm by young women on the other. "If they [country girls] are dissatisfied with the conditions of their lives," warned Crow, "they will have the daring to go forth also, following their brothers, and to take up some industrial fortune in the city whither the bright star of independence beckons."

In books written immediately after World War I, such as Mary Meek Atkeson's *The Woman on the Farm*, the mass movement of the country girl cityward seemed already to be *fait accompli*. "Many thousands of country children in the schools today will be the city dwellers of the future," Atkeson, a farmer's daughter turned PhD, wrote. "And we wish to send them in without handicaps, with the same background of culture, so that they will have an even chance in the race of life." But in this so-called race of life, a handful of postwar rural child-rearing manuals intimated, the country girl was already twice handicapped—first by her birthright, second by her gender. The rhetorical position in which authors such as Crow and Atkeson found themselves proved an especially delicate one—torn between galvanizing the rural population they had grown up with and acknowledging new demographic realities that threatened to forever compromise a heritage they deeply valued. Atkeson walked the gauntlet by focusing on the farm girl's nativity, which, while it might be altered or assimilated, could never be, if properly inculcated, completely lost. The country child, Atkeson enthused, should be imbued with a sentimental and practical understanding of her home place. Just as "the city child learns about the city streets, the municipal government, [and] fire protection" all country girls should know "something of all the many [rural] subjects."

While Crow mostly followed Atkeson's middle road, she devoted an entire chapter, entitled "The Other Side," to the disgruntled, often laconic farm daughters who had grown into manifestations of a "deadly blight." She quoted the sour grapes of frightfully circumspect girls who vowed to the author, "We were crazy to live on a farm and determined not to fail; but as soon as one problem was solved, another would bob up." The sentiments of these unhappy ruralites Crow mercifully paraphrased, quoting only one country girl at length who responded to the author's prompt in describing her typical day:

> I rise shortly before five o'clock and dress hurriedly. Father is calling me to come and strain the milk and get his breakfast. Go down [to] cellar and strain the milk into pans, set them on a large stone table, and skim the milk for cream for the campers along the lake. Measure out ten to twenty quarts of milk and put them into separate pails to be sent out to customers encamped on the lake. Take cream up stairs and put it in a warm place to ripen for churning. Get breakfast, call the children, and after the others have eaten and the boy has started on his morning delivery, I eat breakfast and clear away the dishes. While sister washes them, I mix bread and set it away to rise. Stir the cream, and then sweep three floors and make five beds. By this time it is nine o'clock. Then there are berries to pick, and vegetables to be got ready for market and I go out to help till about half-past ten, when I come

in and make three or four pies and a cake or a pudding. While these
are baking I clean the vegetables for dinner and put them on to cook,
set the table and put the dinner on, meanwhile watching the baking
pies, the rising bread, and the ripening cream. In the course of the
morning ten or a dozen persons have come in for milk, eggs, butter,
or something else, and I have to wait on them and keep their accounts
up in my book. After dinner the bread is ready to make into loaves
and is then set to rise again before baking. While the bread is rising
I scald out the churn and rinse with cold water and then put in the
cream and churn it by hand. After the butter has come and gathered,
I remove it from the churn, rinse the buttermilk out and work the butter; salt and work again and set it in the cellar till the next day, when it
must be worked again and put into pails or jars. Then I pour the buttermilk from the churn into a jar and set it away for future use, clean
and scald the churn, setting it out in the sunshine to dry. By this time
the bread is ready to bake and must be watched rather closely and the
wood fire also. I begin to get things ready for supper, going out into
the garden to pick berries, gather vegetables, dig potatoes, etc. Mean-
time I wait on more people. After straining milk and skimming other
milk, I eat supper and then measure out milk for evening delivery, get
vegetables and bread ready to be delivered also and start the boy on
delivery. Wash dishes and meanwhile wait on milk customers who are
transients. When boy returns from delivery, I wash milk cans and put
them out in the air, write up books of accounts, plan out next day's
work, make list of groceries, etc., that must be bought to replenish our
slender stock. By this time it is ten o'clock; I am weary and my hair is
a sight. After taking off a little of the dirt with a sponge in the wash
basin I tumble wearily into bed until the next morning.

"An account like this arouses a perfect hornet's nest of question marks,"
wrote Crow, further paraphrasing the down-in-the-mouth responses of
other country girls likewise viewed as lacking for books, social outlets, recreation, and fair compensation for their labors. "She has no enjoyments,
no encouragement," Crow lamented on behalf of her hard-pressed farm-girl
readers. "She is hard at work all the time. She neither knows why anyone
would find the farm attractive nor why one should desire to leave it. Time
and interest for her have ceased." How many of these nowhere girls existed,

Crow wondered. "If diligent search is made for them they are found upon the most remote farms where no newspapers penetrate, where the roads are bad and the neighbors far away or beyond forbidding hills, where the deadly round of dishwashing or the weight of work too heavy for the years of the girl are exhausting her strength, stifling her exuberance, and deadening all the power of expression she may have been capable of having." While oppressed and enslaved girls appeared in full view of the public in the nation's cities, the country girl seemed beyond the power of ameliorative uplift, beyond technocratic and academic outreach. This, perhaps as much as any alleged oppression, is what frightened Progressive-era government bureaucrats into action.

Crow's dichotomy proved too neat for the tastes of the many diehard ruralists who distrusted city meddling. Not all farm girls were Cinderellas locked away while their city sisters attended the ball, they pointed out. Such "civilizing" forces as roads, neighbors, newspapers, and mail-order catalogues, did not come without strings attached, and certainly did not constitute the only means by which a civilization might measure its worth. Early in *The American Country Girl*, Crow prescribed the following conditions for "wholly prosperous farmsteads," and yet to some country readers even the outward beneficence of such aims smacked of cosmopolitan ideologies whose implications seemed to work against, rather than for, rural people. Crow's prescriptions for uplift included:

1. "An unrestricted view of the scenery from the living room window" (*Presumption: rural life could only be as worthwhile as it was aesthetically pleasing.*)

2. "A public reading room within reach of buggy wheels" (*Presumption: the town library ought to be regarded as a cultural lifeline.*)

3. "Not so many miles from an accessible city" (*Presumption: rural life was only suitable if moderated by a close town or metropolis.*)

A farm girlhood that lacked these features, Crow believed, would result in "heterogeneous drudgery." The farm girl whose life failed to meet such measures might well find herself in desperate limbo, "destitute of initiative" except for a poorly defined urge to leave. Such an unfortunate farm girl, wrote Crow:

Keenly . . . feels the lack of recreation. She comes to believe that if she were in the city she would not have such late hours of labor. She does not see the twelve and fourteen hour days of work in that rosy dream of good wages and leisured evenings in town. On the farm it is from five in the morning till nine at night; the work is not only too heavy for her, but it is closely confining. She has not the strength for it; and the enforced toil exhausts her energy prematurely. She now sees that the methods used in her household workshop are laborious and out of date; her task is unnecessarily difficult; and who can blame her if under such circumstances her enthusiasm for her work fades away? There is resentment in the remark of the young girl who said: "If we always have to work in an awkward kitchen with rusty old pans, if we do not go anywhere and never have any company, we do certainly want to leave the farm."

Where the pleasures of farm girlhoods were concerned, the bottom line, Crow reminded, was well-suitedness. When a farm daughter migrated to the city, it need not be regarded, she suggested, as a retreat or a defeat, but might in some cases be considered the "call of the best and noblest part of the soul" heard by those who felt a "passion of education, for free access to libraries, [and] for association with intellectual people."

WHILE THE INTENTIONS behind the Country Life Movement were mostly benign, the academic spotlight it shone on the rigors of rural life heightened the notion then widely endorsed by urbanites that farm life was dangerous, if not downright scary—an ages-old bugaboo manifest in films from 1972's *Deliverance* to 1974's *Texas Chainsaw Massacre*, to the 2009 made-for-television remake of Stephen King's *Children of the Corn*, to a more recent horror among the corn, 2011's *Husk*.

In 1866, one year after the Civil War ended at Appomattox Courthouse and eleven years after my ancestor Levi Pickert arrived by train at Davenport, Iowa, an anonymous letter to the editor appeared in the *Delaware County (IA) Union*. The letter had been penned by an indignant farmer's wife subsequent to an impugning of her class by a town girl called "Maggie May." "I have no desire to change my position," the anonymous farmer's daughter wrote, thoroughly miffed, "and would say to all young ladies they may do far worse than to marry a farmer." To the joys and pleasures of the supposedly dreary farm life, the unknown editorialist likewise spoke: "I have another source of pleasure on a farm I could not have in town—the livestock, from the tiny chickens to the great horse. I love to go out after sunset and smooth the sleek sides of the cows and speak to them; and have the sheep come up and put up their pretty faces for a pat and pleasant words; and watching the cunning lambs frolic, which the finest town lady could not help laughing at."

In 1881 another daughter of agriculture defended her life choice in the *Jackson County (IA) Sentinel.* After conceding the many difficult duties incumbent to her position—"hard hours of labor"—she nevertheless avowed, "With all the toil of a farm life, I would not exchange it for a life in town with double the income." In the last paragraph the rhetoric of this Victorian-era farm girl grew still more arch. "Now, how many rosy lassies in the home circle are going to volunteer to rise at five o'clock in the morning from now until the first day of June next, and prepare breakfast with her own hands while the mother naps?"

A generation later the joie de vivre of the Jazz Age made the question still more explicit—what kind of woman would willingly sacrifice the Charleston and the hot jazz of Louis Armstrong to stay at home on her acres? Contributing a letter to the cause of Mary Meek Atkeson's 1924 monograph *The Woman on the Farm*, a nameless Wisconsin farmer's daughter answered the question righteously enough, declaring, "We, the farm women, are where we belong. . . . This close-to-nature elemental existence is the fullest, richest source of emotional satisfaction. And we pay the price. We give our services, and we give up the superficial, sensational stimulants of society and fashion, so we feel square and honest as we take our gifts."

From the Civil War to the Great Depression, newspapers small and large reported with something approaching relish the rugged passion play of frontier and farm justice in what was then considered the West. In 1875 the *Philadelphia Telegraph* reported under the title "A Pennsylvania Tragedy" the story of Middle American farmer's daughter Mary Stokes Russell, who was shot by her lover as he attempted to defend himself from her father's murderous rage before taking his own life. Life was rough and tumble on farms in the great interior, and city and town reportage of tribulations suffered by the nation's farm daughters and wives swung between the sensationally violent and the tragically laconic. In 1911 the *Carroll Times* of Carroll, Iowa, picked up a dateline from Wilton, where a report had been filed under the heading "Flames Leap to Girl's Face." The pitiless blurb read: "The face of Miss Fannie Hudler, a farmer's daughter living near Wilton, was disfigured for life when a kettle full of lard which she was rendering caught fire and the flames flew up into her face. She was to have been married in a few days."

By the turn of the century the argument over the virtues and liabilities of the farmer's wife and the farmer's daughter had been longstanding and vociferous in American history, and more often than not the debate centered on Iowa. In the late 1920s a murder case from the Wisconsin-Iowa border headlined newspapers from Waterloo to Wausau. The grisly AP story opened, "Clara Olson, farmer's daughter, found slain and buried last week on a hillside 15 miles from her home, beseeched Albert Olson, father of her lover, Erdman, last August to make the boy marry her before she became a mother." Less than a decade later the November 28, 1936 post-Thanksgiving edition of the *Mason City (IA) Globe-Gazette* picked up the AP's "Week in Iowa," headlined by two attempted homicides on the farm, the first a Cain and Abel conflict that saw Emmett Patterson shoot his brother Alva because Alva had "tried to run things." The second if-it-bleeds-it-leads story concerned one Melvin Anderson, age 20, who lived on a farm near Webster City and who "went with" Margaret Hess, a 16-year-old farmer's daughter. The AP narrated the near-tragedy in nutshell: "Turned away from her door by Margaret's mother, Melvin returned an hour later, thrust a rifle barrel through the living room window, fired at Margaret as she leaped behind a stove, and scurried from the room on her hands and knees. Telling his sweetheart's father he would kill himself, Melvin fled, but officers finally found him milking a cow at his sister's farm near Sac City."

By the 1950s the image of the farmer's daughter had evolved once again, this time into a more sanitized version that once more lionized the virtues inherent in her rural life, while insinuating that she, like the American housewife more generally, had transcended the early violence, oppression, and backbreaking labor of her foremothers. Mostly it amounted to triumphalist rhetoric, consistent with the postwar boom and America's perceived agricultural superiority. Responding to a recent issue of the magazine *Farm* detailing the improved lot of women in the hinterlands, the *Globe-Gazette* ran an editorial entitled "Modern Farmwife's Life Not All Work, No Play." The opinion piece began by citing the old saw, "The farmer works from sun to sun, the farm wife's work is never done," appending to it this caveat: "The life of the rural homemaker is not the drudgery that it once was." The reason, the editors claimed, was "modern equipment" that removed "some of the work from off the women's hands."

In "The Body and the Earth" Kentucky horse farmer Wendell Berry criticized the bombastic rhetoric directed at agrarian wives and daughters in the years following World War II. "They [women] bought labor-saving devices which worked, as most modern machines have tended to work, to devalue or replace the skills of those who used them," Berry observed. "They bought manufactured food, which did likewise. They bought any product that offered to lighten the burdens of housework, to be 'kind to hands,' or to endear one to one's husband. . . . Thus housewifery, once a complex discipline acknowledged to be one of the bases of culture and economy, was reduced to the exercise of mere purchasing power." America, made self-conscious about the hard, some said criminal, farm labors it had recently asked of its farm women, had set out to make the country life more congenial, a rural version of suburban ease and prosperity.

An exemplar of the fraught transformation from Midwest farmer's daughter to suburban "domestic engineer" was Donna Reed, née Donna Belle Mullenger, a midwestern farm girl whose fates were forever changed when her Aunt Mildred offered her young rural niece a place to hang her hat in Los Angeles after Donna Belle's graduation from Denison High School in 1938. As a girl Donna Belle had been sufficiently proud of her farm heritage to write exhaustively of its every nook and cranny to her pen pal Violet from Pittsburgh. In her introductory letter, dated September 23, 1934, 13-year-old Donna Belle wrote what amounted to a crop report: "The weather is beginning to get cold now, and has frosted a few nights. We certainly did not have much grain to harvest this year, in fact we had none. Our corn we are cutting for fodder." In her letters to Violet, Donna Belle's conspicuous points of pride were consistently farm daughter bona fides—ponies to ride, brothers and sisters to play with, creatures big and small, and farm and school social groups aplenty. "I belong to the 4-H club of Crawford County, Iowa, and enjoy it very much," the teenage Donna Belle crowed, offering in a subsequent letter a hand-drawn map of the Mullenger farm that served as her world entire.

When Donna Belle Mullenger was "discovered" by MGM in Los Angeles in 1941, the studio regarded her farm daughter heritage as, at best, a temporary public relations asset. Studio photographers staged the rising star in nearly every setting a girl could imagine—on the steps of a university building wearing oxfords and cradling textbooks; leaning up against a balcony rail with the wind blowing through her corn silk hair; glancing over a shoulder haltingly at the threshold of an evocatively opened door, her gloved hand suspended somewhere between coming and going.

The most arresting of the studio shots captured in 1941 are those publicizing the young ingénue's rural roots. In one the Midwest-bred beauty is shown gathering eggs in a basket; in another, she's captured looking provocatively over her shoulder in rolled up short shorts that border on Daisy Dukes. She wears cowboy boots, her hands on the handles of a single-row planter with a horse hitched in front. The images were only part fiction, as before Donna Belle Mullenger became Donna Reed, she had in fact milked cows and helped in the fields. Indeed, Reed famously made a bet with the Academy Award-winning actor Lionel Barrymore that she could milk a cow, and Barrymore took the young farmer's daughter turned movie starlet up on her boast, dragging her to a nearby dairy farm where she performed as promised and collected on the wager. Still, Donna Belle Mullenger, a Midwest farmer's daughter, was bound to be a transient creature, destined to be retired—put out to pasture—as soon as MGM's initial publicity blitz was over. In her place was "Donna Reed," a studio-coined sobriquet Donna Belle would learn of secondhand while leafing through an industry news magazine in 1941; she instantly disliked it, though she waited until 1976, reported biographer Jay Fultz, to confess that she found it "cold" and "forbidding." "I hear 'Donna Reed,'" the actress said, "and I think of a tall, cool, austere blonde who is not me."

Circa 1958's new TV season, Donna Reed was just 36 years old, but Hollywood, and the rest of America, appeared to have lost its taste for 1941's humble girl straight from the farm or even 1946's self-sacrificial Mary Hatch in Frank Capra's feel-good classic *It's A Wonderful Life*. Donna Reed, Academy Award-winning actress, was finding it difficult to land a single meaningful movie role in the mid-1950s, when, with her then-husband Tony Owen, she launched her own self-titled situational comedy, *The Donna Reed Show*. Just as Donna Belle Mullenger, earnest 4-H farm girl, had been transformed into a World War II pinup girl, she now morphed into an upper-middle-class housewife of Hillsdale, Donna Stone. Reed dutifully wore her character's aprons and checkered dresses until 1966, when she could stomach the straightjacket no longer.

In retrospect Reed's many transfigurations—some elective, some required—closely mirrored the changes experienced by farm-reared women more generally in postwar America. Perhaps the iconoclastic Wendell Berry had a point—that it wasn't the working woman, but the woman who worked at home in a domestic economy, be it farm or household, that America so desperately wished to distance itself from after 1960, making for a contested ground where the locus for the culture wars was no longer *if* a woman worked, but *for whom* she worked.

Responding to a firestorm of criticism whipped up when, in a piece for *Harper's* magazine, Berry let slip that his wife typed his manuscripts, the author answered his detractors, "It is easy enough to see why women came to object to the roll of Blondie, a mostly decorative custodian of a degraded, consumptive modern household, preoccupied with clothes, shopping, gossip, and outwitting her husband. But are we to assume one may fittingly cease to be Blondie by becoming Dagwood?" Farm wives, he hastened to add in "Feminism, the Body, and the Machine," who voluntarily contributed their labors to a self-sustaining household economy free and clear of corporate exploitation, were, he lamented, "apt to be asked by feminists, and with great condescension, 'But what do you *do?*'" This slight Berry regarded as symptomatic of a systemic vilification of precisely the kind of household economies farm daughters like Donna Belle Mullenger had been born into and had increasingly decided to leave. "What are we to say," Berry asked, "of the diversely skilled country housewife who now bores the same six holes day after day on an assembly line. What higher form of womanhood or humanity is she evolving toward?"

IN JUNE OF 1950, one month before my newborn father was examined and declared fit by our family doctor, Dr. Littig, in rural Mechanicsville, Iowa, *Farm* magazine featured on its cover Mary Furleigh, daughter of Mr. and Mrs. Robert Furleigh of Clear Lake, Iowa. The picture showed Mary playing piano at a tea held in connection with the YWCA rally, and bespoke the kind of urbane, cultured freedom to which it was believed rural women of the day aspired.

Fifties farm exchanges like this one were arranged between Old World and New in order to share cutting-edge agricultural methods and technologies, to develop potential overseas markets, and to trumpet just how enlightened and "civilized" America's farm women had become. Soviet premier Nikita Khrushchev toured the US in 1959, famously visiting the Roswell Garst farm in Coon Rapids, Iowa, and came away from the trip most impressed with the gleaming, touch-of-a-button kitchens enjoyed by American women. The women of the Iowa Farm Bureau, meanwhile, met in their state chapters and raised $10 in each of ninety-nine counties to bankroll Lucie Boeak's and Frau Maria Baur's visit to a Hawkeye State farm. When the two fraus reached Mason City, they, like Khrushchev, marveled at the relative conveniences available to America's farm women. "Farm women in Germany," the *Globe-Gazette* reported, "help with the field work from dawn to sunset, much of the farm work being done by hand. In addition, they keep house, cook meals and rear children."

Though my own farm mother was forbidden as a child and young adult from doing the heavy farmwork and unwelcome in the kitchen, she, like many of the farm daughters of her generation, learned, at least symbolically, the lessons of manual labor. "In the summer when there was corn to cut out of the beans," she recalls, "mom would line all five of us up in the field and we would each take two rows. We were young, but we had these great big sickles. Here were these little kids with sickles hopping through the beans. But no one ever got cut. No one ever got hurt. . . . We'd go in the evening, and we'd do it probably two or two and a half hours, and we got 50 cents an hour."

My aunt Barbara on my father's side tells a similar story—not so much of pitching in to help with chores necessary for survival, but of occasional labor assigned to introduce to her and to her sisters the concept of chores. "On the farm I was in my own little world much of the time . . . playing on the swing set, in the sand pile, playing with the kitties," Barb recalls. "Watching Daddy do the farmwork was a perk of being a country kid. . . . I could ride a round or two on the tractor with him, help him feed the pigs, and watch him work in the shop. Many, many times I followed him around the barnyard while he did his nightly chores. Town kids didn't get to spend that much time with their dads."

"We all had to help pick garden produce and strawberries," Barb adds, "and help to preserve or can them. [Daddy] always said that girls should not do farmwork, so we never had outdoor chores. We did occasionally have to walk the beans and pull button weeds." Conversely, Barb reports, her brother, Michael, had "a lot of expectations placed on him," as he was expected to follow in his father's footsteps. "His role on the farm was much harder than ours," Barb admits, "because he didn't have anyone to share the burden with." While her father believed in separate spheres for farm boys and girls, the bottom line, Barb tells me, is that he wanted his children to be happy. "He wanted each of us to be married and have our own families," she says. "I believe he expected [his son] to take over the farm, but would not have stood in the way of Michael pursuing other dreams. A farmer wants to pass the farm to his son."

Barb's eldest sister, my aunt Patricia, tells a similar story of a coming of age on a midwestern farm in the 1940s. "Working was just what we did," she tells me. "We were born with the idea that when things needed to be done, they got done. I can't say I disliked anything. And we didn't lack for anything. I was pretty introverted, so if I had books, magazines, paper— we wrote on both sides—and a big box of crayons, I was happy. I probably wouldn't have been able to be self-employed later in life if it hadn't been for my farm upbringing."

The eldest, Patricia benefited most directly from the presence of my great-grandmother, Amber, "a force to be reckoned with," who lived on the same property. "She was full of energy and busy all the time. She got up and did the chores and made breakfast before she woke the rest of the family. She taught me to sew, do needlework, cook, and she could tell the best stories. She had plenty of time for her first grandchild." The most bookish of the three farm girls, Patricia, not surprisingly, liked school the best, and was the sole daughter to attend college, first in Des Moines and later in Iowa City, after her daughter Mindy was born. The only one of the sisters to live her entire adult life in the country, Patricia attributes her departure from the more urban life of her siblings mostly to her choice of marriage partner, Charlie, a small-town laborer from just down the high-way, and, to a lesser extent, to her own chosen career as a seamstress that lent itself to small-town entrepreneurship. "Escorted to the alter by her fa-ther," read Patricia's July 28, 1967 wedding announcement in the *Cedar Rapids Gazette*, "the bride wore a black street-length evening dress with a white moiré jacket with pearl buttons." The dress, the article noted, "was designed and made by the bride." Unlike her sisters' husbands, profession-als who worked in nearby Cedar Rapids, Patricia's mate was a tradesman, a lineman at Iowa Electric Light and Power Company whose work often took him into rural areas. Her sisters left for town, Patricia maintains, not because they were necessarily eager to get away, but because that's where their husbands' jobs took them. "I was happy with it because I liked the country and thought it was the best place to raise the kids," Patricia re-calls of the first years of what would become thirty-five consecutive years of rural living after she and Charlie built their country home in 1976.

IN JUNE PROFESSOR of history David Danbom agrees to meet me to talk farm-girl economics at Atomic Coffee on Fargo's historic Broadway on a day of torrential summer rains. After nearly forty years of teaching at North Dakota State, Danbom's in the process of moving from the town he's called home since 1974, and, as we settle down in the back room of the Atomic, he's content to let the moving boxes in the garage wait a few hours while we plumb the history and mysteries of the farmer's daugh-ter. Danbom has trained himself to see rural America through the eyes of a social realist rather than a romanticist, and his trademark sobriety explains why I have driven hundreds of miles to seek his opinion. If any-one could cut through the mythos of the farmer's daughter, I reasoned, it would be him.

When I ask Danbom to give me a nutshell history of the farmer's daughter, he begins on a blue note. "There wasn't much in the way of mobility. The only occupation really that was available to many farmers' daughters was schoolteaching . . . in part because school patrons saw them as more appropriate for teaching young children, but largely because they were cheaper than men." Beyond teaching, he says, the farm daughter's alternative was to marry, and most teachers, he explains, did marry after a short time. "To marry meant becoming a farm wife and replicating the farmer daughter experience with a new generation, or being a spinster, which was considered quite unattractive," he summarizes. "But eventually you had migration of young women to cities, really beginning after the Civil War in the late nineteenth century when you have more urban occupations open to women, secretarial occupations, which at the time were overwhelmingly male. Secretarial schools and simple things like the typewriter made a big difference, women's fine motor skills being better than men's on balance."

As each born-on-the-farm generation grows into motherhood or fatherhood, it exercises a far-reaching choice—whether a next-generation daughter is fated to carry the label *farmer's daughter* or a son to become the no-less important *farmer's son*. Dropping the label by leaving the life, Danbom endeavors to tell me in this, his *Reader's Digest* history of an icon, is what tens of thousands of American rural women began to do in the late 1800s. "You have the rise of corporations and white-collar work. Pink-collar work, as we call it now, became available, and so you begin to have this draining off of population. People have other alternatives to staying on the farm. In a place like Fargo we had business schools where farm girls would come to train on stenography and typing, and then would get jobs in banks and insurance companies and businesses."

The allure of secretarial jobs in the city attracted the most gifted farmers' daughters from across the Corn Belt, ushering in an era when those same transplanted daughters of agriculture would meet, marry, and often raise their children in the nearest industrial city rather than at home on their rural route. My great-aunt Helen, for example, whom my great-grandfather once praised in his book *The Furrow and Us* as the "leading lady" of the farm, found secretarial work 60 miles away from her family's farm with the John Deere Company in the early 1940s. By 1944 this young, beautiful, highly competent farmer's daughter was already upwardly mobile. "I regret very much that the bearer, Miss Helen Jack, has requested a release from our organization, but I can highly recommend her to anyone wishing the services of an efficient secretary or stenographer," Deere Company's M. A. Prince wrote on behalf of my great-aunt in a letter dated February 29, 1944. "She has had experience in all types of office work. She takes dictation rapidly and her typing is always exact and correct. You will find her very conscientious, pleasant, and always on the job."

In the Depression, claims Danbom, the population of Fargo would have been just under thirty thousand and on average at least six hundred of those farm-to-town émigrés would have worked as live-in maids. As in Moline, fresh-off-the-farm women would often pursue maid-work for a few months to acquire a feel for the city and its opportunities, then move up to other jobs in restaurants, laundries, hotels, or businesses, turning the cooking, cleaning, and bookkeeping skills they had picked up as farmers' daughters into ready cash and, ultimately, promotion. Then along came World War II, which upset the carefully ordered rural Midwest culture as no war before or since. Farm boys coming home from "Gay Paree," as the old song goes, were less likely to marry farm girls after having seen the world, and more likely to marry someone they met during their travels—a German or Japanese war bride perhaps—or a young woman who had struck their fancy upon their return to port. "I think World War II is really a huge break. It's a huge disjuncture," my interviewee tells me. "The number of farms declines, the rural population declines. The traditional role of farmers' daughters and farmers' wives changed."

Historically, midwestern women contributed to the family income through the production of cream or butter and eggs and poultry, but after World War II, Danbom explains, new sanitation rules made it difficult for small dairy farmers to operate. Poultry, too, went the way of mega producers like Tyson. "So women who want to contribute to the family income and who need to contribute to the family income have to look somewhere else to do that. One of the good things that happens after World War II is that towns and cities become much more accessible to women, so . . . you can live on the farm and drive there and work at the Ben Franklin store or Walmart or teach school or work as a nurse and in that way contribute to the economic situation of the family while in many cases finding a reward-ing social experience."

As rural population precipitously declined, the entire cultural fabric woven by generations of country women began to unravel as it became harder and harder to maintain the farmers' and homemakers' clubs and the churches and women's auxiliaries, the complex organizational struc-tures Danbom calls the "intricate social underpinning" of rural life. "Farm-ers' daughters are part of a family system," he reminds. "One of the inter-esting things about a farm is that the home is also the business. Today it is one of the last of those places. Here in Fargo in 1930 there were nearly one hundred grocery stores in town and in most of them the family lived above or behind the store. So in that case a home was also a business. Ag-riculture was where this model was really embedded. And in that sense farmers' daughters played a role like farmers' wives and farmers' sons in maintaining and perpetuating the family business. . . . Farm women had an intimate interest in it in a way that maybe the wife of a retail merchant didn't have."

Danbom confesses he wishes more attention might be paid to the pio-neering rural-to-urban female migrants of the late nineteenth to early twen-tieth centuries, and to the role they played in the rise and development of American business, because the farmer's daughter of that watershed mo-ment of rapid urbanization had, as my interviewee puts it, the "skills and language" many immigrants did not. "If you want to look at male immi-grants, they're absolutely important in mining coal and making steel and slaughtering beef and doing all the other muscle work in the late nineteenth and early twentieth centuries," he confides. "But if you look at office work, retail work, which is really more important for the future economy of the United States, that was largely done in those first couple generations by farmers' daughters. Economically that's important." In Danbom's view it's no surprise that the exodus of farm girls in the early twentieth century cre-ated a national "demographic crisis." Men, he says, were bound to follow wherever enterprising women went. "When women leave one of two things is going to happen. Either men are going to follow, or men are going stay there, and they're not going to find wives. Reproducing the family system then becomes difficult."

Where the loss of the family business is concerned, Danbom resists easy nostalgia. "Historically a farmer's daughter was kind of a subset of a view of farmers as sort of innocent, unsophisticated yokels who could be manipulated by city people. With farmers' daughters it was always thought that they were very innocent and could be manipulated sexually. This is the whole genre of farmer's daughter jokes, which, even when I was a kid, you heard all the time. I don't know when the last time I heard a farmer's daughter joke was. There was that *Seinfeld* episode . . . that's the last I even heard a reference to it." Danbom's referring to the episode called "The Michigan Bottle Deposit" where Newman is taken in by a red-and-white-plaid clad farmer's daughter and her bib-wearing dad, whose only condition is "keep your hands off my daughter."

In his book *Born in the Country* Danbom gently piques those who, where they once decried the alleged exploitation of the farm girl and wife, now refuse to ascribe a higher rationality to her leaving. "Some feminist scholars argue that women who were removed from farm work were trivialized, disempowered, and diminished in status," Danbom writes. "Some women," he points out, "were empowered by farm work; but for others, scouring milk cans and slaughtering chickens was drudgery, and many farm women today find that their off-the-farm work adds new dimension to their lives. Most urban women who work outside the home find satisfaction and even fulfillment in their jobs; is there any reason to think that rural women are different?"

PRIOR TO MY search for an American icon I'd assumed that farm daughters left the farm mostly due to the ceaseless and often thankless work, until I'd been confronted with the fact that both sets of grandparents in my own long-standing Midwest farm family had believed daughters should be excluded from heavy-duty choring. "We had to do the dishes at a very young age . . . one of us girls would dry, one would wash, and one would put away," my mom tells me, recalling her childhood responsibilities on the farm. "It wasn't really bad until the summer when the hay balers came over, and then we had to produce a full meal for about twelve people.

It would take all morning to prepare all the food. After everybody ate their feast, we ate what was leftover—we spent about two hours cleaning up the dishes and the mess. And then we were expected to clean our rooms on a regular basis. . . . We didn't do any livestock. We were expected to work in the garden, which of course we hated. We were either helping pick strawberries, or picking beans, or picking peas and shelling peas." The girls never learned to can, possibly because their mother was a perfectionist who would rather do the job right even if it meant doing it herself.

Coming of age in the late 1950s and early 1960s, none of the girls milked cows, though they did get the hands-on experience of caring for animals, even if it was, by my mom's own admission, a "naive" undertaking. "Every spring we would get these orphan lambs," she recollects. "If we fed them and took care of them, when we sold them we would get the money to go to camp. That was our exchange. So we'd feed them with big Pepsi bottles with a great big black rubber nipple, and we'd mix up the formula, and we'd go there twice a day and feed the lambs, which was great fun. . . . We had all the usual names, Midnight, Smokey, that sort of thing. . . . Had we ever known that ours was blood money, I don't think that we would have gone to camp. We got them plump, and they went off to be sold. We never thought about who they went to or why they wanted them."

But what my baby boomer mother and her sisters missed out on in unpaid chores, they made up for with compensated employment. With four siblings, my mom got her share of involuntary child care experience, though her chance to parlay that practice into real wages came off the farm, when, at age 12, she took her first job babysitting for 25 cents an hour and later, at age 15, began caregiving for the elderly at the local nursing home. If my mother and aunts, Heartland farm girls all, are any indicator, the celebrated work of a farm girlhood had already begun to decline in practice by the 1940s, and had pretty much heard its death knell by the Sexual Revolution of the 1960s, when my mom herself married a Midwest farmer's son mere weeks after her high school graduation. Of my four aunts on my mother's side, exactly half hitched their wagons to farmers, aiming to be the next generation in a bucolic tradition. A generation later, all were living in town, having separated from or divorced their husbands, or having kept the husband but lost the farm.

What was lost in the arena of economics and ethos, I want to know, when a generation of daughters left the farm or stayed on without a true working share in its economic survival. "Just that hardworking ethic," my mom offers. "If something doesn't work, you persevere. We depended on each other as best we could. In some ways that's a downfall of farmers, too. They don't reach out. They're independent. . . . So what is lost? I would say the can-do, figure-it-out-for-yourself attitude. That's not to say that kids that live in the city and towns don't have that, but when it snows in town, in two hours you know the plow is coming by. If it snows on the farm, you're going to be snowed in for two days, and what are you going to do to entertain yourself? That's essentially the difference."

I opt to check my mom's generational hypothesis against the generation X farmers' daughters I know, all born 1971 or later, beginning with my friend Becky Kreutner, an elementary schoolteacher. Kreutner grew up on a farm in Eden Township in Benton County, Iowa, where her dad farmed 200 acres. Her father, she tells me, always said that farming was a wonderful way to live, but a terrible way to make a living. Kreutner describes a prototypical farm childhood surrounded by family as neighbors. "Some years where leaner," she remembers. "But we always had food on our table, clothes on our back, and lots of love to go around." I ask Kreutner, whose favorite T-shirt as a kid read "Anything boys can do girls can do better!" to recount some of her own memories of chores and choring. "When I was in high school I helped a neighbor and that neighbor's grandfather unload several loads of straw. There was also a boy about my age helping. The grandfather was so impressed with how I outworked him that he sent me a card, which I have to this day. It reads, 'They found something in Vinton, Iowa, that does the work of five men . . . one woman!' He signed it, 'Thanks for the help. I appreciated it.'" Kreutner remembers a mother who worked right alongside her father as an equal partner in the operation. "Dad discussed everything with my mom and all decisions were made together," she says.

Kreutner, who describes herself as a tomboy that, underneath her seed-corn cap, was frequently mistaken for a boy, says she loved being outdoors and helping her father with daily chores and other seasonal jobs such as walking beans, baling hay, vaccinating piglets, and moving hogs. Her sister, meanwhile, gravitated more toward the domestic tasks like fixing meals and doing laundry. "I grew up to have a deep respect for those of the opposite sex that hold the same values and work ethic as my dad, brother, and others I knew," she tells me. "I realize that women do not possess the same amount of raw strength as men, but they can work side by side with them, and their thoughts and opinions should be valued."

Next on my generation X list is Angela Crock, a lab technician who grew up a farmer's daughter on a 200-acre farrow-to-finish hog farm east of Oxford Junction, Iowa, where her family grew corn and soybeans in rotation. When as a girl Crock talked her father into buying a horse, they began making hay was well. At the family homestead in the 1980s schoolwork partially supplanted farmwork as the chore of a farmer's daughter and her contribution to the family, so much so that Crock routinely completed her homework in the farrowing house.

For Crock chores solidified and codified larger familial and societal roles. "We always square baled our hay and each one of us had our specialty," she recalls. "My sister always drove the tractor pulling the baler, while my dad and I stacked. My mom made sure we had lots of cold water, and she helped put the hay in the barn when we brought it in from the fields. It was a family affair. I think the best part of being a girl growing up on a farm was that it taught me so much about responsibility and family values. I learned that at a very early age—I think even earlier than my 'city folk' counterparts. We did everything together, from working on the farm to having fun on the farm." Crock's first pet was a pig named Radar who could hear her coming with a bucket of feed long before any of the other pigs. Crock recollects, "Although she was in a pen with other sows, I was responsible for taking care of them. I also spent many hours up at night bottle-feeding piglets rejected by their mothers. I was responsible for beings other than myself, and I think that is a huge lesson in responsibility."

In my interviews with successive generations of Midwest farmers' daughters, however, a disconnect emerges between symbolic or aesthetic chores and choring for economic survival. The chore memories of farm girls born 1970 or after are invariably bound up with sentiment. At the same time that Crock mentions feeding hogs, for instance, she describes "hours riding on the trails that my dad had cut through the woods," enjoying the "wildlife as well as the foliage," "the smell of leaves in the fall," and "watching the large population of deer and turkey from horseback." Chores, for many country girls in my sister's and mother's generation, contained an elective element, the chorer often permitted to choose, within limits, the tasks that best suited her or served her wider education. For baby boomers and generation Xers chores frequently entailed the luxury of specialization that had gradually infused even farm labors since the turn-of-the-century, when an annual report of the New York State Agricultural Society observed, "In these days of specialists, the farmer's daughter must be one, too."

When Crow wrote of the country girl's contribution to the American zeitgeist, citing a "great rural reserve of initiating force, sane judgment and spiritual drive" and a "clearness of eye," she saw such virtues and habits of mind arising from the farm daughter's number one calling card: hard work. She wrote:

> In thousands of farmsteads, they are helping their mothers wash dishes three times a day three hundred and sixty-five days in the year. . . . They are carrying heavy pails of spring water into the house and throwing big dishpans full of wastewater, regardless of the strain in the small of the back. They are picking berries and canning them for the home table in the winter; they are raising tomatoes and canning them for the market; they are managing the younger children; they are baking and sewing and reading and singing; they are caring for chickens and for bees and for orphan lambs; they ride the rake and the disc plow and sometimes join the round-up on the range.

Well into the 1930 and 1940s a farmer's daughter was defined largely by the work she performed in service to the family industry. In the 1920s agricultural scholar Mary Meek Atkeson printed an excerpt from a letter she received from an anonymous farm woman, reiterating Atkeson's point that the farm woman "learns from experience that the greatest joy and contentment come from hard work at worthwhile tasks." Atkeson's farm woman correspondent concurred, writing, "There is comfort, real comfort, in plain living . . . in the sweet old-fashioned way of working till you are tired. . . . In tending little lambs and chickens and calves, in watching things grows, in making your own delicious butter, in picking the raspberries from your own patch, and in getting the first peas and sweet corn, and in the fall in gathering the apples from the trees that your own great-grandfather planted on the farm years ago."

That the farmwork shouldered by the letter-writer and by my great-grandmother and grandmother alike differed in both breadth and depth, style and substance, than that completed by my sister and mother and aunts, suggests one reading in the changing character and identity of farm daughters and the region they helped define. "Being a farm woman was a full-time job," my aunt Patricia remembers of the farm women in our clan who preceded her, those that in the late 1940s taught her the skills that would make her future career as a self-employed tailor possible. In addition to doing outside work, my great-grandmother Amber somehow found time to sew, do needlework, cook, and serve as raconteur. Other family accounts have her killing, plucking, and cooking chickens, and baking incomparable pies, often in the same day. In 1912 a teenaged Amber boasted to the other farm girls and boys in her letter circle, writing, "Hay was a good crop this year. We put 150 loads in the barns. I drove the horses to the hay fork to put it in the mows. My little brother, 8 years old, drove to the hay loader one whole day, and he had three horses to drive. It was quite a trick to manage them. Oats are all in the chock ready for the threshing machine. A great many people are threshing now, and we will thresh next week."

That in the public imagination the farm daughter's calling card has changed over the course of a century from tireless laborer, to aesthetical, ideological, or sexual creature parallels a larger trend from the less-than-sexy demands of working farms to the more aesthetic pleasures of hobby farming on exurban acreages. But as the experience of the region's great-grandmothers and grandmothers suggests, the challenge of choring may never have been antithetical to the pleasures of farming. As the half century of rural history intervening attests, to be a farm daughter or son "saved" from the work of the farm was, in a sense, to forfeit family farms that worked.

CHAPTER SIX

WELK GIRLS
AND DAISY DUKES

THE DAY AFTER Thanksgiving in 1950 my grandmother's family, the Puffers, made headlines in the *Cedar Rapids Gazette*. Writ large across the top of page was a picture of a bustling family of six staring intently at a Cold War–era box with its space-age screen. The set was switched off, but my relatives—Uncle Donny and his wife Edith, my great-grandparents, and young cousins Phil and Steve Puffer—dutifully posed as if glued to the tube. Under the image of the captivated Middle American family, the caption read, "For farm families, whom both work and weather may confine to their rural homes during the winter, TV will unfold like a magic carpet. Here, the Puffer family, living two miles southwest of Mechanicsville, gathers around its set."

It's a picture of contrasts. There's the family patriarch, Big Daddy, clad in overalls, reclining in an overstuffed chair in the middle of a traditional midwestern living room with its wood floors and rails, its floral wallpaper and Persian rugs, newspapers and magazines stacked high beside him, a remnant of a print era soon to be eclipsed. On the photo opposite my clan's in the *Gazette*, Mrs. M. J. Barnes, the wife of the Vinton, Iowa, mayor, was shown gathering her four children in front of the set, including what the caption described as "the fascinated boy" of the family, Hughie. Below the Barnes clan was another snapshot, this one showing a family in front of a big wood-trimmed console; the only one looking at the camera was the dog, which the newspaper captionist wryly noted had little interest in the "new-fangled contraption."

"In ever-increasing numbers, eastern Iowans are buying TV sets," the *Gazette* reported. "Among the hundreds of TV set owners and watchers in eastern Iowa, those on this page are perhaps representative of the farm and city groups who are scanning the screens." Eastern Iowa farm families like mine had just three principal stations to choose from in 1950, the year my father was born—WHBF-TV, Rock Island; WOC-TV, Davenport; and WOI-TV, Ames. Just ten years earlier Cedar Rapids enjoyed its first real taste of a mass-market media frenzy when Major Edward Bowes announced he would salute the city to a Columbia Network audience numbering 35 million. When the city learned that no locals were scheduled to appear on the program, Charles "Pud" Moel, a lyric tenor who lived at 1518 North Street Northwest in the old cereal town, was hastily summoned and shipped off to New York for an audition. A breathless *Gazette* article announced, "60 of the fastest telephone operators available in the vicinity would be hired to tabulate Cedar Rapids votes for favorite amateurs." While Moel didn't win the "Amateur Hour" that Thursday in 1940, the mayor declared "Major Bowes" day in Cedar Rapids. The Hawkeye State had cast its media vote for one of its own, an underdog in every sense, and had relished every mass media minute of it.

In 1956 the Czech town on the banks of the Cedar River lucked out again with news that a new network TV sensation, ABC's Lawrence Welk, would make Cedar Rapids a stop on a two-week tour of one-night bandstands. The up-and-coming farmer's son from Strasburg, North Dakota, had married Fern Renner at the Cathedral of the Epiphany in Sioux City, and Iowans remembered him fondly for it. A congratulatory ad taken out by Cedar Rapids ABC affiliate KCRG in 1955 saluted: "The Lawrence Welk

Show has brought an unusual avalanche of mail from KCRG-TV viewers of all ages praising it on behalf of the entire family." The show aired at 7 p.m. on Saturday nights, by which time many hard-working Heartland farm families had finished their day's tasks and were ready to relax. A Midwest farmer's son, Lawrence Welk was a Saturday night institution across the Corn Belt. My mom, whose own German mother and Czech father counted themselves among Welk's entourage of farm family boosters, looked forward to Saturday night as a welcome break in what she, coming of age in mid-1960s, considered farm tedium. "Saturday night, if we had a fun night, came closest to family time," my mom recalls. "If Dad was home and Lawrence Walk was on, we'd all gather around the TV set. We would watch and eat popcorn."

By the time Welk and his Champagne Lady arrived in September of 1956, Cedar Rapidians were starstruck. "The show presents the large and versatile Welk aggregation with his dancing and singing 'Champagne Lady,' Alice Lon," the bill for that Saturday in August read. The young bandleader had been voted top dance band of 1955 by *Downbeat* magazine, and the TV musical show of the year in a poll by *Radio-Television Daily*. Nine thousand people jammed the parking lots at Hawkeye Downs Speedway on the south side of town in what the *Gazette* called "fine football weather" but "considerably short of fine for an outdoor concert."

Critics dismissed Welk's folksy Middle Americana as schmaltz, but those who assembled that chilly night disagreed emphatically when the whistle-worthy Lon appeared in a low-cut bouffant better suited to summer. The Lennon Sisters, recently discovered by Welk's son, made the trip as well, impressing the reviewer with their "combination of youthful prettiness and lovely voices," making them "as delightful an act as has ever been seen in Cedar Rapids." Welk invited anyone in the audience to dance with him that night, his Champagne Lady, or anyone else in the band, and the wait was lengthy. "If anyone had the slightest doubt about the popularity of the Lawrence Welk Orchestra," the *Gazette* huzzahed, "it was dispelled fast and emphatically."

One year after doing the polka with his Hawkeye State fan base, farmer's son Lawrence Welk achieved a Nielsen Rating of 30, a mark that translated into an unbelievable third of the American TV-viewing audience.

"His success as a TV entertainer . . . still seems a little unbelievable," an incredulous Larry Wolters wrote in the *Chicago Tribune* in 1956, "and it appears even more improbable in light of his background." The urbane Wolters pointed to the musician's hardscrabble upbringing on a North Dakota farm as an obstacle to, rather than as an explanation of, the bandleader's fame. Indeed, Welk's personal story was one many rural Middle American families could relate to, a story of unlikely immigrants settling an inhospitable land, spending their first winter under the shelter of an overturned wagon covered in sod, or so the Welk legend went.

As with many celebrities and VIPs who grew up before World War I, an epic American home place figured prominently in a larger-than-life story. "Out where the prairie meets the sky in the broadest of arcs, and the lazy clouds drift slowly over the fertile earth," Mary Lewis Coakley's serial biography of the star opened in the *Chicago Tribune*, "lie the billowing fields and the sod house. The place holds many clues to Lawrence Welk's character and to his success." In her multi-installment piece Coakley relayed a story near to fairy tale: the accordion the maestro learned to play on the farm—one of just a couple of family heirlooms the Welks brought with them from Alsace-Lorraine—had been passed through the family by way of a blind ancestor; Welk had nearly died from an appendix burst in the barnyard; too sick to attend school he had recouped by practicing on the $400 mail-order squeezebox his father had promised him in return for the proceeds of his son's farm labors until Welk turned 21.

"Out there on the Plains, we were really apart from the rest of the world, so most everybody around just spoke their native tongue," Welk recalled for his biographer, explaining how not just the geography and weather had shaped him, but also the Ursuline nuns who taught him reading, writing, and arithmetic. "I guess we kids didn't really think that out at the time, but some of it rubbed off on us, and it taught us a lot—about unselfishness, and unworldliness," the bandleader recalled of the farm-daughter and son- culture in which he grew up. On the farm Welk hauled wheat from the threshing machine to the bins, often playing weddings gigs until 1 a.m., then waking up at 4 a.m. during harvest and threshing time to complete his chores. "Our parents taught us that life was meant to have hard conditions," the accordionist recalled. "How else could we grow strong?"

Welk's story, including tales of having to shovel his way to the barn in 30 degree below zero temperatures, resonated with Heartlanders, in particular, and solidified his down-home image. For those who tuned in by radio or television in the Corn Belt listening to Welk equated to a moral choice—a vote for the small-town and rural midwestern values they might reasonably hope to pass on to their daughters in contrast to what seemed the lurid sensationalism emanating from Hollywood. Welk pitched his squeaky-clean image primarily to parents, which in rural America meant showcasing wholesome young women of the kind epitomized by the farmer's daughter. Here too the North Dakotan bandleader aired on the conservative side, picking acts that bristled with the kind of wide-eyed, Middle American energy Welk himself had brought to Chicago and Los Angeles in the years before his show hit the big time. In 1958 Welk's program was anchored by Lon and the Lennon Sisters, the latter of whom the *Chicago Tribune* feted in a feature article entitled "Lennon Sisters Are Average Girls," wherein writer Joan Beck observed, "It's hard to find a less sophisticated teenage foursome. None of the Lennon Sisters has ever gone steady, although Dianne, Peggy, and Cathy date classmates from St. Monica's parochial school. And no boy is permitted to take any of the girls out until their father has called him aside for a man-to-man talk about safe driving."

The Lennon Sisters weren't literally Midwest farmer's daughters, but they might as well have been. Their father, Bill, had worked for a dairy before his daughters became icons, and like Welk back on the farm playing rustic squeezebox behind the barn, the girls possessed the requisite lack of refinement, as none could read music. Bill described the Lennon home in California as nearer in spirit to Laura Ingalls Wilder than the industrializing Midwest farms of the day, commenting, "Singing always has been a natural part of our family life, just as household chores. Whenever one of the babies needed changing, Dianne or Peggy always did it without waiting to be told." Like the stereotype of the pitchfork-wielding patriarch protecting his progeny from would-be usurpers, Lennon predictably frowned on rock and roll's suggestive lyrics. Though they sometimes fell into a rock and roll rhythm, the Lennon Sisters' trademark numbers, including such farm-friendly fare as "How You Gonna Keep 'Em Down on the Farm," "A Bushel and a Peck," and "The Lonely Goatherd," boosted the foursome's appeal as fresh-faced milkmaids.

Alice Lon likewise traded on her representation as a small-town girl, the kind for which Middle American moms and dads as well as sisters and brothers would fall. "In the course of his TV hour Mr. Welk dispenses a most generous helping of what the moderns in his music dismiss as plain 'corn,'" wrote Jack Gould for the *New York Times* in 1956, adding that a big part of the appeal was "the usual attractive feminine vocalist Miss Alice Lon." Like Welk accordionist Myron Floren, who had grown up on a farm in Roslyn, South Dakota, Lon had come from a rural place called Delta County, Texas, and as a would-be paragon of Middle American virtue, she appeared in full skirts and petticoats designed by her mother. Welk publicists highlighted Lon's all-American roots from the get-go, pointing out that she had married a former football star and that, despite doing two shows a week on ABC, she still had plenty of hours free for the family. In a 1958 article that ran in the *Chicago Tribune* under the title "The Champagne Lady Bakes a Cake," Lon avowed, "Next to singing and dancing, I'd say baking cakes and cookies is one of my favorite activities. And it's more fun than ever, now that I have three young sons and a handsome husband to request such numbers as chocolate cake or sugar cookies." Included lest there be any doubt about her American girl bona fides was Lon's recipe for Quick Cocoa Cake.

Lon epitomized the Welk girl until 1959 when she was sacked, according to the Museum of Broadcast Communications, for "showing too much knee" on camera. Though he received thousands of letters of protest and later tried to hire Lon back, Welk's actions demonstrated that he meant what he said when it came to the wholesome image expected of his regular female singers. The six-line Associated Press blurb seemed likewise verklempt on the subject of Lon's firing, reporting, "Alice Lon, a singer known as the Champagne Lady, has left the Lawrence Welk Orchestra. Mr. Welk said it was because of a disagreement over choice of songs."

The press spin on the surprise firing seemed to reflect a Middle American bias toward conflict avoidance. When it came to media criticism, midwesterners ultimately sided with Welk. He had listened to them, brought respectable young women into their living rooms, and they would reward him with their lifelong loyalty. By 1955 when the Welk show began as a summer replacement for ABC, Welk had barnstormed for a dozen years around the Midwest, making countless allies along the way. Like a Populist politician he had hugged and danced with the girlchildren of thousands of Heartland families, including my aunt Patricia, who, as a girl, once left my grandfather's arms for a dance with North Dakota's most popular farmer's son. In Cedar Rapids the editors at the *Gazette* posed the rhetorical question du jour, "Who are his fans? Just people. Millions of them. People who like sweet music and music in which they can recognize the melody. People who sit in their homes each Saturday night and relax for an hour with ABC-TV feeling a pleasant link between themselves and what has aptly can be described as the Welk family."

In the same year, 1956, the *New York Times's* Gould concurred, remarking, "Mr. Welk's arrangements are often 'sweet' in the extreme, but they are also clean-cut and varied. Unquestionably they will be an anathema to many; indeed, they have been summed up as 'Mickey Mouse Music.' But for others who are content merely to hum a pleasant tune, their appeal would seem self-evident." Self-evident enough, as it turned out, that when newspaper and magazine critics began to take Welk to task for sentimentality and pandering, letters in his defense arrived by the bushel. When Larry Wolters accused Welk and other entertainers of being sloppy dressers in a 1958 *Chicago Tribune* column, angry listeners filled the newspaper's "TV Mailbag." "Lawrence Welk is one of the best groomed men on TV. Don't sling mud," reader Harry G. Allen tut-tutted, while Mary Jones warned, "You are making a mistake saying these uncomplimentary things. Larry Wolters, you can't be feeling very good. Better sleep around the clock." Another reader, choosing to remain nameless, fired back in Welk's defense, "Concerning sloppy dressers, let's have your picture on *TV Week*, and I'll give you my opinion of you."

By 1961, when my 11-year-old mother, her parents, and her four siblings watched Lawrence Welk religiously each and every Saturday night on the farm, Welk had begun to symbolize a growing cultural rift between the "greatest generation" and their mostly teenaged baby boomer children, a schism that found itself reflected in the politics of the Sexual Revolution. While in a five-year primetime run Welk had trounced his competitors, headlines in the *New York Times* as early as September of 1961 proclaimed, "Lawrence Welk to Vary Format of TV Show to Meet Competition." Reporter Richard Shephard noted, "Lawrence Welk, facing keen Saturday night competition from network movies and Westerns on other channels, is planning some variations"—variations that would include a series of themed shows. A 1964 Associated Press story entitled "Welk Has Simple Recipe for Success" conceded, "The avant-garde musicians and teenagers may find the Welk school a bit square, but he is not interested in their reactions." The article quoted the bandleader's explanation of his strategy for pleasing audiences with "the kind of entertainment that should come into the home." For his part Welk unapologetically pinpointed his target demographic. "The ladies are my number one audience," he told the AP's Cynthia Lowry, "and the gentlemen are number two. We also have about 5 million young children in our audience. We are short on teenagers, but they are a problem. I can't afford to go after a teenage audience and lose my regular one. I don't even try."

Back in the Midwest rural families like mine were beginning to register a generational split in their viewing habits, one that would have profound implications for farm daughters and sons. In my paternal grandparents' household the family still watched Welk, mostly in deference to my grandfather, but the children, including my 14-year-old father, were more enthused when rock and rollers like the Beatles and Elvis appeared on *The Ed Sullivan Show*. Increasingly, seeing Cissy and Bobby do "How You Gonna Keep 'Em Down on the Farm" dressed in straw hats and checkered plaids seemed quaint even to real-life farmers' daughters and sons like my aunts and uncles.

In her memoir *The Horizontal World* North Dakota farmer's daughter Debra Marquart recalls the cultural moment in the early 1960s when Welk and his orchestra began to seem more and more like an anachronism: "On the Welk Show it appeared to me that no creative demons were being exercised," she writes, "no addictions fought, no lust succumbed to or overcome, no one was struggling against the forces of censorship or racism. . . . The Welk musical family seemed too aware of the camera, too eager to entertain me." Later Marquart recalls, by way of contrast, the night the Beatles played *The Ed Sullivan Show*, remembering how the mop tops so irked her farm father that he dismissed himself to go to the bathroom, and stayed there until the lads from Liverpool had finished their caterwauling.

AFTER ABC CANCELED Lawrence Welk in 1971, the show remained popular in syndication at over two hundred stations, but was viewed as "too old" to appeal to advertisers. ABC's strategic move away from aging Middle Americans in the Rust Belt and Corn Belt was closely and more famously followed by CBS in a "Rural Purge" that would leave country families without network TV representation for the first time in generations. Much of urban America had already moved from rural stereotypes to rural indifference, so much so that in 1971, the year of my sister's birth, CBS axed *Hee Haw* along with rural-themed shows *The Beverly Hillbillies*, *Mayberry R. F. D.*, and *Green Acres*. In doing so they cited network judgments that the shows reflected the "wrong" demographics (i.e., rural, older, and less affluent) even if the programs remained popular in the ratings and with the people. The mastermind behind the cuts was CBS's Fred Silverman, who instigated the so-called Rural Purge after market research in the late 1960s showed advertisers attracted to younger, more urban audiences. For its part rural America didn't know what to call the elimination of its favorite programming—the term "Rural Purge" hadn't yet caught on—it simply knew the rebuff hurt. "Mr. Silverman's agenda was explicit," Edward Hoyt of *The Washington Post* later wrote. "Whether it was a strict business maneuver based on smaller advertiser demands for those show's large audiences or Mr. Silverman's personal bias, a cultural sting was associated with the purge and is still recalled bitterly." Indeed, the popularity of the axed shows suggests the purge of so-called "Hayseed TV" was not, in fact, a dollars and cents decision at all, as *The Lawrence Welk Show* would last in first-run syndication for eleven years and *Hee Haw* a whopping twenty-two. In farming areas like mine, country dwellers would now have to make do with highly caricatured, highly sanitized versions of themselves if they hoped to see their rural brethren on the boob tube.

Writing about the end of "corn-pone" television, the *Post*'s Paul Farhi agrees that place, rather than politics or business, dictated the change in tastes. "For the past three decades," he writes, "network television has generally eliminated depictions of regions we've come to think of as 'red': southern, Midwestern, mountainous, rural, exurban." TV dramas, he adds,

have for two generations "extolled blue-city living, and marginalized, condescended to, or simply ignored just about everywhere else." Farhi points
out what most in my grandparents' generation well knew: rural America
had dominated prime time as recently as the 1950s, when shows such as
Cheyenne; *Gunsmoke*; *The Restless Gun*; *Have Gun, Will Travel*; *Wagon Train*;
Rawhide; and *Bonanza* ruled the television and topped the ratings. In the
1960s the pattern largely held, the Hayseed TV boom shifting the emphasis
toward small-town and rural comedy in high-performing shows like *The Beverly Hillbillies*; *The Andy Griffith Show*; *Mayberry R. F. D.*; *Green Acres*; *Gomer
Pyle U. S. M. C.*; and *Petticoat Junction*.

On any given Friday in the spring of 1965, in the blissful TV years
before the purge, eastern and central Iowa families like mine could open
their *Iowa TV Magazine* supplement from the *Des Moines Register* and find
there half a dozen rural programs running in their broadcast area, including, in the morning, *The Andy Griffith Show* and *The Real McCoys*, and
in the afternoon, *The Rifleman*, *Trailmasters*, and *The Huckleberry Hound
Show*, a program that featured a blue-haired, folksy dog with a fondness
for song. In the evenings they could cozy up to *Gomer Pyle U. S. M. C.*,
Rawhide, and *The Farmer's Daughter*, where for 101 episodes from 1963
to 1966 a jaw-dropping Inger Stevens played Minnesota farm girl turned
big-city nanny Katy Holstrum. Stevens's show would likewise fall under
the network guillotine in April of 1966; four years later its star would
commit suicide.

BY THE TIME *The Dukes of Hazard* came along in 1979, Middle
America had agitated for the years following the Rural Purge for a show
it could call its own, one wherein country sons and daughters played starring roles. While urban critics panned *Dukes* early and often for its rural
stereotypes and backwards, backwater portrayals, these, at least, were characters rural America knew. Boss Hogg, for example, wasn't unlike many of
the small-town, small-minded bankers Heartland farmers found themselves
swindled by, the very type they would strike out against in the depths of an
all-too-real agricultural crisis. Sheriff Rosco seemed to my cousins and me

a pretty close match for our community's sheriff, whose underlying good nature seemingly left him stupefied in pursuit of the most petty crime. Bo and Luke Duke were just like us, we surmised, obsessed with muscle cars and fast motorcycles and kicking up dust on country roads. And of course there was Daisy, the beautiful, clever, down-to-earth farmer's daughter reminiscent of the neighbor girls who lived within hollering distance of our Heritage Farm.

Already by the spring of 1979 *Dukes* was a phenomenon, the most popular among the eighteen replacement series that had gone on the air that winter to attract thirty million viewers. Predictably, critics raised high brows at the show's down-home ethos. In the pages of the *Chicago Tribune* Lee Winfrey joked, "[T]his is a series where the characters race cars more often than they change clothes, stirring billows of dust while they raise many whoops of 'Yahoo!'" One month later, with *Dukes* still tearing up the Nielsen Ratings, bulldog *Tribune* critic Gary Deeb set out to sink the upstart rural show among Chicagoland's urban and suburban viewers. "*The Dukes of Hazzard* is a good-ol'-boy romp that borrows shamelessly from *Smokey and the Bandit* and *Gator*, a pair of stupid but popular Burt Reynolds movies," Deeb sneered. "The show also employs every . . . stereotype you've ever seen, heard about or imagined." In spite of critics' attempts to ground the General Lee, however, by January of 1980 *Dukes* stood alone atop the Nielsen Ratings, beating *Dallas* and *60 Minutes* with a Welk-worthy score of 30.5.

While for us boys the red-orange 1969 Dodge Charger the Dukes drove down Hazzard County's dirt roads was the real star, the actors earned our respect, too. Bo Duke, played by John Schneider, hailed from Mount Kisco, New York, a town like our county seat, Tipton, Iowa, with a population of less than ten thousand. His partner, the shorter, darker Tom Wopat, who played Luke Duke, seemed even more like us, the rising star born and raised in Lodi, Wisconsin, a typical Midwest corn town of a couple thousand souls. Of the two male leads it was Schneider, 24 years old in 1981, who captured most of the media attention with his 6-foot-4 frame and sandy blond locks. While critics conceded his potential to be a latter-day matinee idol, they condescended to his character, on screen and off. A December 1981 *Tribune* article reduced Bo Duke to "a country bumpkin who races cars with the abandon of a kamikaze pilot" and Schneider himself to a righteous, old-fashioned yet ambitious Middle American type who believed in "happy endings and old-fashioned values." "Drugs and booze and easy ladies are not his style," the *Tribune* smirked, painting Schneider as a joyless teetotaler and killjoy fundamentalist, albeit a good-looking one.

But it was farmer's daughter Daisy Duke, more than Bo, Luke, or even the General Lee, that turned the collective heads of a nation. Catherine Bach's portrayal of Daisy squared with the popular image of the farmer's daughter as a fiercely independent, divinely loyal, undeniably beautiful young woman caught in the middle between boorish locals and big-city dreams. Even though Bach could not have been better cast—she had grown up on a ranch outside Rapid City, South Dakota—critics predictably panned Daisy's character. "Another Duke on the show is another cousin, Daisy, impersonated by Catherine Bach," Lee Winfrey poison penned. "Miss Bach is a poor actress who talks as if she is reading words off a blackboard. But she is so gorgeous none of that matters."

Feminists, in particular, objected vociferously to Daisy's short shorts, though they failed to point out her strengths: loyalty, resourcefulness, and guile. While Daisy often invokes feminine wiles to help her cousins and her Uncle Jesse, the sultry bedroom scenes of *Dallas* and *Dynasty* were altogether absent in *Dukes*. Indeed, beyond her short shorts and provocatively batted eyelashes, Bach's Daisy turned out to be an unusually independent female for 1970s sitcom television, though few urban critics acknowledged as much. That Bach's character waited tables at the Boar's Nest and stayed at home on the farm to help her cousins out of tight jams so rankled urban ideologues that Bach was left to point out the obvious to the incredulous: Daisy Duke really was a lot like the young women she grew up with in the rural Midwest.

"My mode of dress was never been attacked by the Moral Majority," Bach recalled in a 1981 interview with United Press International's Kenneth R. Clark. "It's been attacked by feminists maybe. . . . I've had feminist journalists who have said, 'Don't you get upset with the way they shoot you?' I was a little upset with wearing shorts myself at first . . . and they said, 'Go and see what the girls are wearing at the restaurant across the street.' They were all wearing little mini-skirts that matched the table cloths!" In fact, the real-life story Bach shared with the *New York Times* in 1982 seemed to come straight from farm daughter central casting. Bach's ranching father wanted her to stay close to home after she graduated high school in Rapid City, just as Donna Reed's dad had wanted his pride and joy to attend the University of Iowa. But Bach, like Reed, had a sense of wider worlds and greater destinies. Mere weeks after graduation she took out all her money from savings, and like Reed before her, headed west to Los Angeles for drama school. "She is a friend of mine," Bach said of the Daisy Duke character that made her a household name. "A very good friend. She is part of my life."

Bach brought a superabundance of Midwest country girl traits to her *Dukes* audition in 1974, including the sewing of her own clothes. "I made all my clothes the first season, in response to a really horrible wardrobe choice by the producers," Bach told *The Canton (OH) Repository* in 2009. "They wanted me to wear a red-and-white checked poodle skirt, a white turtleneck, white go-go boots and—get this—a blonde wig. The executive producer guy had a big crush on Dolly Parton, and I think she wore that on one of her album covers." As for the car obsession that served as a cultural mainstay in Hazzard County, Bach confirmed that she had lived that, too. "It's really cold in the winter in South Dakota and one of the things people do is work on their motorcycle or a special car inside the heated garage. So I just grew up with that car culture," Bach told reporter Dan Kane. While critics on the Coasts begrudged Bach her success, the response in the People's House proved closer to the go-get-'em-girl reaction registered by many in Middle America. In October of 1981, with the show at the height of its popularity, Bach dropped off her famous Daisy Duke poster at the White House for her old teacher Shirley Moore. Moore, the *Lakeland Ledger* reported, wrote her former pupil a thank-you note that gushed, "I'm the envy of the White House, and I'm having your lovely picture framed. Mrs. Reagan saw the picture and fell in love with it."

Beyond Daisy Duke, however, on-screen farmers' daughters were fast disappearing from Heartland televisions. In 1982 *The Lawrence Welk Show* ended its first-run syndication the same year Schneider and Wopat threatened to sue Warner over a contract dispute and left *Dukes*. Fans groaned at the plot contrivance when two look-alike nephews of Uncle Jesse, Coy and Vance, replaced the absent Schneider and Wopat on the set; were we really supposed to believe that Bo and Luke had suddenly left Hazzard for the NASCAR circuit? The original Duke Boys did return for the 1983 season, but the damage was done. By February of 1985 the last vestige of prime-time rural television had left the air, and with it the last remaining portrayal of a starring farmer's daughter.

A generation later a *Dukes* resurgence was well underway as 1980s TV kids began to grow nostalgic about the television of their youth and sought to share those wonder years with their children. In Nashville Duke Fest drew upward of 150,000, while reruns of the show on Country Music Television drew 23 million viewers. In 2005 a whole new generation of movie-goers ogled Daisy—played by Jessica Simpson—in a feature film that cast '70s heart-throb Burt Reynolds as Boss Hogg and Willie Nelson as Uncle Jesse in a re-visionist, socially conscious plot pitting the Duke boys against the dastardly coal industry. Online in the new millennium Bach hawked her own line of Daisy Dukes, priced upwards of $50 for the retro feel-goods that came along with the assurance of all-American cotton picked in the US.

Though much of *Dukes'* enduring popularity can be chalked up to ad-oration of the beautifully cunning Daisy, Bach's website biography proves anxious to recast her as an "actress, activist, and businesswoman," in much the same way that historians have felt the need to reinterpret Donna Reed's farm-based biography for postmodern political correctness. Bach, the website boasts, "has transcended the character that made her a household name." Still, the brisk online trade in Daisy Duke retro gear suggests that the country girl Bach played way back when remains her most enduring legacy. On Internet auction sites it's not Bach's jewelry or jeans that are hot on the auction block, but Daisy Duke "adult costumes" consisting of denim minis with built-in petticoats and off-the-shoulder, tie-front blouses. Daisy's principal commercial artifact on eBay—a caricatured costume suitable mainly for donning at masquerades and naughty nights out—suggests the complex appeal of her image among generation X fans.

As has often been the case in popular culture, America may have attempted to heal its own hegemonic guilt over the debasing of an indige-nous culture—in this case the farm daughter—first by caricaturing it, then by laughing at it, and finally by co-opting its dress in ironic homage. But lost in such two-faced tributes is one important fact: the farmers' daughters I knew growing up really did wear halter tops and jean shorts, not to increase their tips or get their kicks, but for thrift. Our daughters of the farm wore their Daisy Dukes on the most sizzling summer days, not to be looked at or to work hard at being someone they were not, but simply to work hard. On the farm, frayed jeans were routinely and proudly repurposed as serviceable cut-offs for farm boys and girls alike. For all of Daisy Duke's imperfections as a role model, the very idea that a farmer's daughter who lived with her family and paid the bills by working a tough job surrounded by tough locals could be down-home in denim instead of surreptitious in sequin, was in its way, a qualified victory for the country girl as portrayed in popular culture. While the beautiful elites of *Dynasty* and *Dallas* appeared more often in evening gowns and cocktail dresses, hatching plots and spinning webs, in Daisy Duke Midwest farmers' daughters had found, if not a working-class heroine, a woman who could at least be counted on to look out for herself.

CHAPTER SEVEN

MILKMAIDS IN MANHATTAN

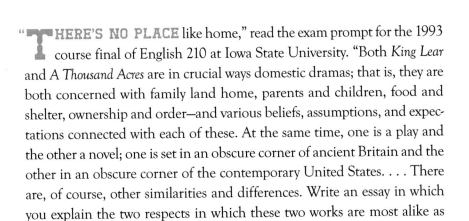

"THERE'S NO PLACE like home," read the exam prompt for the 1993 course final of English 210 at Iowa State University. "Both *King Lear* and *A Thousand Acres* are in crucial ways domestic dramas; that is, they are both concerned with family land home, parents and children, food and shelter, ownership and order—and various beliefs, assumptions, and expectations connected with each of these. At the same time, one is a play and the other a novel; one is set in an obscure corner of ancient Britain and the other in an obscure corner of the contemporary United States. . . . There are, of course, other similarities and differences. Write an essay in which you explain the two respects in which these two works are most alike as domestic dramas . . ."

I remember tearing into the exam, ink pen flying across the wide-ruled pages of my mandatory exam blue book. I wrote with something approaching pride, pleased by the plum, topical draw of a contemporary Midwest writer. It didn't seem to matter that I didn't share my professor's enthusiasm for the book or that Jane Smiley was a native of Los Angeles who had attended Vassar and was thus about as far away from a Heartland farm girl as she could get without falling into the Pacific.

Indeed, New York publishers and literati have long held a fascination for the farmer's daughter, a mostly parasitic infatuation evolved over time to reflect the changing tastes of urbanizing generations. In 1941 the *Mason City (IA) Globe-Gazette* and other Midwest newspapers picked up a new serial fiction by Allen Eppes—author of such female-minded pulps as *Eveless Eden, The Glass Slipper, Go West Young Maid, Southern Belle,* and *Strictly Feminine.* Syndicated by the Central Press Association, the series ran under the title *A Maid in Manhattan,* and told the story of one Susan Farmer, a duty-bound farmer's daughter living with a spinster aunt. Taken at face value the themes resembled those of Smiley's *A Thousand Acres*—dutiful farm women, obligation to family, conflict between the microcosmic world of the barnyard and larger societal forces—but with a campy, allegoric twist.

By the 1940s the farmer's daughter had already become a novelty for much of America. Opening November 7, 1941, the first episode of *A Maid in Manhattan* introduced readers to college-educated leading lady, Susan Farmer, a farm daughter only to the extent that she lives on a failing farm with her aunt Alice. Her father died years earlier in an equestrian accident, and she, still in mourning and wanting not to abandon her aunt, has thus far resisted the many marriage proposals made to her by her boyfriend, Fred. Meanwhile, back in Manhattan, up-and-coming ad man Roy Leonard has drummed up a can't-miss ad campaign for his client, the perfectly alliterative Dainty Diana Dairies, by which the company will sponsor a "Typical Farmer's Daughter Contest," with the winner to receive $5,000 and a trip to New York City. In the car with the amorous Fred, Susan hears the radio spot Roy has dreamed up: "Send your picture to Dainty Diana Dairies, in care of this station. That is, if you feel you can qualify as the Typical Farmer's Daughter for whom we have been searching. Also send us a letter of not more than 250 words telling something of your life on the farm." Fred snickers at Susan's interest, commenting to his best girl, "You look more like something straight from Park Avenue than you do something straight from a farm. You never even milked a cow in your life." Susan Farmer, however, sees in the contest a chance to save her family's farm, and thus demands of her boyfriend, "I want you to snap me in a sunbonnet, holding a milk pail and standing beside Esmeralda." Esmeralda is Aunt Alice's last remaining cow.

Back in the Big Apple, Roy wears a "satisfied smile" at the novelty of the genius promotional idea that has already landed him an office in New York's tallest skyscraper. "God bless all the dear little daughters of farmers!" he chuckles. "He owed a lot to them, bless their innocent hearts. And boy, how they had occupied his thoughts here of late! He had been thinking about them for weeks!" The roots of Roy's promotional campaign, as it turns out, are sentimental, originating with a rural girl—Lucy Bell, Mary Anne, Betty Lou—he can't remember exactly which—that he had fallen in love with at the age of sixteen. But mostly his objective is commercial: "to put the Dainty Diana Dairies products before millions and millions of American housewives."

Chapter six, which arrived on November 13, 1941, found Roy pouring over contestant photos. He was vexed, for "not by the wildest stretch of the imagination could he visualize a single one of them within a mile and a half of a bovine mammal," that is, until he espies the picture of a girl who makes "his heart climb up into this throat and try to do a high dive." "The girl was standing with a milk pail in one hand," Roy notes, "and the other resting on the rear-end of a somewhat forlorn looking cow. Skin like those peaches his grandmother had been so proud of, hair like corn silk in the summer sun, trim ankles, slender body, and wearing a gingham frock and a sunbonnet pushed back and from her smooth white forehead."

"Gosh all hemlock!" Roy exclaims at the sight of the girl he feared extinct, and all of a sudden he is awash in memories of "Lucy Belle . . . Betty Anne . . . Mary Lou . . . all rolled into one. That summer he was sixteen. His grandmother's farm. Dreams. Hopes." Roy takes the picture straight to his boss, Mr. Daniel Van Wert Jones, who, gazing longingly upon the picture, declares they've found their winner along with a new visage to adorn their dairy products. Mrs. Jones, however, isn't so sure about the ethics of the contest, tut-tutting her husband for "bringing an innocent girl like this one to New York . . . [and] subjecting her to the evils of a big city."

Whether Middle Americans in 1941 were honestly taken in by the convenient myths of *A Maid in Manhattan* is impossible to say, but what is certain is that the good folks of the Heartland eagerly read the newspaper novel for forty-six consecutive episodes. While Susan Famer turns out to have been voted the most sophisticated girl in her college class, in addition to speaking French, Italian, and Spanish, she qualifies, by birth anyway, as a farmer's daughter. The episodes also confirm that she is everything Roy believes her to be—short of a girl who actually milks cows and plants fields as her staged snapshot insinuates. When, in the end, Roy, not Fred, proposes to the sweet, vivacious, resourceful Susan Farmer, and she accepts, midwesterners could be happy, seeing a grandson of a farmer who'd turned into a Manhattan ad man tie the knot with a sophisticated, clever, yet down-to-earth girl who cared enough about her family's farm to stage a bogus contest entry to pay off its debts.

In popular fiction and drama up and through the 1980s, the farmer's daughter was generally portrayed in a positive light. Usually she found herself protected from serious critical scrutiny by the allegorical tales in which she appeared. In *A Maid in Manhattan*, for instance, Susan's surname alone paints her as representative of her class. The treatment is flatly allegorical; her hometown goes unnamed, thus she belongs to Middle America, not to a particular town, city, or section. The serial novel makes no claims of realism, especially where an icon like the farmer's daughter is concerned.

The same pattern held true for much of the farm daughter drama of the era, including Millard Crosby's 1938 one-act *She Was Only a Farmer's Daughter*. In a prefatory note from the playwright, Crosby makes it clear to his audience and actors that his production should be regarded as "sort of old-fashioned," preparing his reader for comedic "melodrama." Crosby provides this telling synopsis of his leading lady: "Born a farmer's daughter, Millie has longed for an education. So she goes to the city where she becomes a victim of that deep-dyed villain, Mulberry Foxhall, the most despicable scoundrel that ever trod the boards." Mulberry, Crosby informs readers, appears "on the scene suddenly, as is the wont of all stage villains, and proceeds to make life intolerable for the farmer's daughter in the most approved villainous manner. The parts are all excellent, and you'll howl with laughter."

Like Susan Farmer, Millie is a parentless orphan looked after by an aunt who worries, as innkeeper Ella Smithers puts it, that "she'll probably marry some worthless pup who'll steal the farm right out from under her nose." As the action opens Millie has lived in the "wicked city" for a year, and when she returns to her village Ella reminds her that she "mustn't ever forget that [she's] a farmer's daughter." To this injunction Millie replies, "As if I'd ever want to forget it. The farm is my real home, and all of the hogs and sheep are my dearest friends. It seems like only yesterday that I was one of them. . . . I'm afraid that city folk are inclined to be hard-hearted and stand-offish."

Emblematic of her type, Millie fits the mold of the farmer's daughter in nonliterary fiction and drama prior to the Sexual Revolution. An orphan, she is twice exceptional not just for her rural upbringing, but also for the especially poetic conditions of her rearing. As her Aunt Sally reminds, "Your cradle was rocked up in the attic afore your Paw mended that leaky roof. I allus said twas the raindrops that made you so beautiful, Millie." Thus, when the urbane, disingenuous Mulberry Foxhall enters, the audience is prepared to recognize him on sight as villain. "Tall and dark, he is suave and polished in manner," stage descriptions remind us. "A large, exaggerated black mustache covers his upper lip and hangs down on each side of his mouth. He wears striped trousers, a cutaway or Prince Albert coat, spats over his highly polished black patent leather shoes, [and] a large flower in the buttonhole of his coat." For good measure and lest Middle American audiences miss his exaggerated evil, Crosby accessorizes his antagonist with a cane, a silk hat, and pair of kid gloves.

REREADING A *Thousand Acres* a decade and a half after my long-ago course final, I am struck by its author's near total repudiation of the Midwest farmer's daughter—struck by how, between daughters Ginny, Rose, and Caroline, Smiley presents us with three almost farcical manifestations of a type deployed in what was supposed to have been a "realistic drama with echoes of King Lear set in Iowa." The farm crisis had exposed rural America and made it vulnerable, and, as an apparent consequence, the fashion in farm literature circa the mid-1980s began to turn from lightly comedic farce, melodrama, and allegory, to darker "realism" to keep pace with what the rest of America believed it was learning about shocking psychodynamics down on the farm.

Smiley's pseudo-realistic representations of farm daughters put rural readers in a fraught place even as those characterizations risked stereotypes far more dangerous than the simple, country girl melodramas of World War II and before. In *A Thousand Acres* farm daughter Ginny is an adulteress, and, but for a botched attempt to poison her sister, a murderer. She is also a middle child, a peacemaker, the type of woman that keeps most Middle American families speaking to one another—someone the author wants readers to recognize and identify with, even though her many vices run the risk of thwarting that affiliation. Smiley attempts to coax our identification with Rose, too, by giving this second sister a disease that perennially plagues the rural Midwest—cancer. Even so Rose is "caustic and determined" as the book's jacket puts it, the rough-around-the-edges farm daughter who calls them like she sees them, refusing to mince words. Additionally, the character of Caroline is drafted to ring a regional bell, likewise representing a Heartland stereotype: the high-flying farmer's daughter who leaves the farm for a white-collar profession—in this case a law career in Des Moines—where she grows distant and unaffected in the face of serious dramas occurring back home.

Sometime in the early 1980s a decree, it seemed, had been issued that any award-winning literary novel written about the agrarian Midwest must not be about prosperous, productive farmers and the verdant fields they plowed in real life, but about deep oppression and loneliness suffered at the hands of its women and, sometimes, its men. Farmers must be stoic to the point of brutality, their wives and daughters long-suffering and fearful.

The first major *New York Times* review of Smiley's epic ran in 1991 under the heading "On an Iowan Farm, a Tragedy with Echoes of Lear." Since the agricultural recession of the 1980s the Heartland had become, as if by some perverse word association game, yoked with the word "tragedy" or "crisis," and *A Thousand Acres* played directly into that zeitgeist. The review penned by Christopher Lehmann-Haupt was a positive one overall, praising the author's "tragedy in a cornfield" and pointing out an underlying anti-agriculture theme that played well among the literati on the Coasts. "In *A Thousand Acres*," the reviewer noted, "much of the evil that is set loose is a result of humans attempting to exert control over nature." It was understood that "attempting to exert control" applied to both the abused female protagonists in Smiley's domestic drama and to the land itself. For Lehmann-Haupt, however, Smiley had gone overboard in her portrayal of the farming father, the Lear character, Larry Cook, the subject of much of the novel's most exaggerated vitriol. "There are moments," observed the *Times* reviewer, "when the results are grating. It seems too much to make Larry Cook a sexual abuser of Ginny and Rose in their childhood. Larry himself is hugely egotistical, but to push his selfishness into pathological monstrosity insults him retroactively and robs him of his majesty." For the many critics on the Coasts already conditioned to look for such conceits, the idea that the story of three Midwest farm daughters would involve sexual abuse, alcoholism, adultery, senility, madness, and attempted murder seemed about par for the course, though Heartland farm families begged to differ. Whatever moral objections Smiley's few salt-of-the-earth readers might summon wouldn't be allowed to matter; the literary establishment had hailed *A Thousand Acres* as a brave work of "daring risks," and, unless outraged rural readers had money enough to buy space in the *New York Times Book Review* or the ear of the board members of the Book Critics Circle in Manhattan, a realistic drama of daring risks it would remain.

By the time the tome captured the Pulitzer Prize for Fiction in 1992, its message had been reduced to a highly gendered critique of domestic abuse in farm country. "Jane Smiley's *A Thousand Acres*, a novel that weaves the plot of King Lear through the story of an Iowa farm woman who was sexually abused by her father, won the Pulitzer Prize for fiction yesterday, the April 8, 1992," the *New York Times* announced. Gone were even the most passing references made by earlier reviews to the book's "portrait of the American farm." The story played better and appeared more prize-worthy to urban book buyers, publicists realized, as a story of victimization set in a rural world widely suspected of oppressing its daughters. Clearly book buyers, the vast majority of whom did not live on Middle American soil, liked what they heard, *A Thousand Acres* debuting on the *New York Times* best-seller list immediately after scoring the Pulitzer. Smiley entered the list at number twelve in the company of such heavyweights as Toni Morrison, Michael Crichton, Sue Grafton, and Dr. Seuss.

In one fell swoop the rural Midwest had landed in the hands of hundreds of thousands of readers nationwide, but *which* rural Midwest—the one I had grown up in, which, though it was touched by such ravages as alcoholism and abuse, wasn't defined by these transgressions but by their opposite, neighborliness and nurture? Smiley had literally imagined for tens of thousands of the farm-uninitiated not just the environmental excesses of the men and women who grew their food, but their alleged psychosexual trespasses as well. To abuse and alcoholism the best-selling authoress had now added adultery, poisoning, blinding, dread disease, and a litany of biblical sins needing several hundred pages to fully explicate.

Increasingly, the book was read not as a clear-eyed portrayal of an independent rural America in need of help, but as a critique of the very ideologies and foundations upon which that society had been founded—not as the story of a proud family faced with the difficulties of inheritance, but as the psychological trauma and drama of a raped and ravished farmer's daughter. Writing for the *Times*, reviewer Ron Carlson boiled the book down to its core questions: "What . . . is it to be a true daughter? And what is the price to be paid for trying one's whole life to please a proud father who only slenderly knows himself—who coveted the land the way he loved his daughters, not wisely, but too well." Writing for *The Washington Post* Writers Group, Ellen Goodman described the book's arc as moving from "a child's ingratitude to a father's abuse." No mention is made of the farmer's economic plight, his conscience, or his ethical and physical struggle to put food on the table.

Smiley's Pulitzer-winning fiction opens with Ginny's recollections of her father, recollections predictably troubled in Freudian fashion. "My earliest memories of him are of being afraid to look him in the eye, to look at him at all," Ginny opines. "He was too big and his voice was too deep. If I had to speak to him, I addressed his overalls, his shirt, his boots. If he lifted me near his face, I shrank away from him. If he kissed me, I endured it, offering a little hug in return."

No surprise, then, that as Ginny recounts the story of the loss of the family's titular thousand acres, Larry Cook, her father, is portrayed as drunk and abusive, her husband, Ty, as timid, ineffectual, and emotionally distant, a poor partner in Ginny's eyes and an even poorer lover. Smiley designs for Ginny, her supposedly prototypical farmer's daughter, a history of five

miscarriages that have precipitated a loss of physical interest in her arche-
typally repressed Middle American husband. In the rare moments that Ty
and Ginny make love, Ginny arouses herself with thoughts of her lover, Jess
Clark, with whom she will have a roll in the hay the following afternoon in
the rough bed of a pickup truck. Smiley doesn't allow either Ginny or Ty
the dignity of lovemaking in the literal sense of that term. Ginny, fantasiz-
ing not of her husband, but of her lover, must picture herself prostrate in
the act, back "humped as a sow's, running in a smooth arc from my root-
ing, low-slung head to my stumpy tail." Neither does Smiley allow Ty the
artistry or ability of foreplay, Ginny describing her husband's animal ad-
vances thusly: "Ty woke up. I was panting, and he was on me in a minute."

Predictably, in Smiley's stereotypical portrayal, the daughter most free
from Ginny's and Rose's suffering is the youngest, the one who managed
to escape the farm's gravity and brutality via a professional career in the
nearest metropolis. Caroline represents the clan's achiever. "If we kept her
at home," Ginny recollects of the baby of the family, "she would languish,
do badly, seem like nothing special." The fourth offstage victim in the story
is Larry's deceased, martyred farm wife, once more straight from Smiley's
stock casting of Midwest female types. She had a history, Ginny reminds
herself of her mother, recalling the many romantic artifacts extant in her
mother's closet from the glory days before she met her tyrannical mate. In
this tomb to a mother's young womanhood, Ginny and her sisters find
seven pairs of high heels and "two cylindrical hatboxes, and in these were
eight or ten hats, some with flowers or fruit, most with half veils" and "four
or five corsages with their pearl-tipped pins stuck into the satin-wrapped
seams." "I don't want to make too much of our mother by asserting that
she was especially beautiful or especially distinguished by her heritage or
her intelligence," Ginny begins her recounting, and yet clearly her mother
is distinguished by her beauty and grace as much as her husband Larry is
distinguished by his loathsomeness.

In a story reported by the Associated Press, Smiley claimed the book
had been inspired by a drive she took from Minneapolis through the Heartland on Interstate 35. In 1991 the Iowa Humanities Board jumped on
that bandwagon, paying the decorated writer handsomely to share with
Iowans her views about their most vital industry. In her speech "A Thou-

sand Acres: How Much is Too Much," Smiley debuted as public agricultural critic, conceding, "We all know that what is happening in Iowa agriculture is compelling and urgent." And yet very early on the delicacy of Smiley's unlikely rhetorical position that night found her disclosing an utter lack of experience with the worlds she had made a literary fortune representing. "My great-great-grandfathers—two of them doctors, one of them a newspaper editor, and one a photographer—were themselves several generations removed from the farm," she confessed in her speech's disclaimer. Smiley concluded her address that October night by quoting to her audience an extended passage written by writer William Kitteridge concerning not the satisfactions of farm life, but, predictably, its pitfalls:

> We could not endure the boredom of our mechanical work, and couldn't hire anyone who cared enough to do it right. We baited the coyotes with 1080, and rodents destroyed our alfalfa; we sprayed weeds with 2-4-D Ethyl and Malathion, and Parathion for clover mite, and we shortened our own lives. In quite an actual way we had come to victory in artistry of our playground warfare against all that was naturally alive in our native home. We had reinvented our valley according to the most persuasive ideal given us by our culture, and we ended with a landscape organized like a machine for growing crops and fattening cattle, a machine that creaked a little louder each year, a dreamland gone wrong.

A dreamland gone wrong? In its way Smiley's speech had been brave, though, significantly, she delivered her address to the ideologically simpatico Iowa Humanities Board and not to the Iowa Farm Bureau. Much as she had done in her novel, her address conjured a world that was not hers by birthright, that she did not intrinsically empathize with or know—one, in fact, that the text of her speech shows she found thoroughly toxic if not morally untenable. Those listening close to the evening's commissioned talk were likely to have heard the author's foreshadowing of her exit strategy, her plan for leaving the very farms and farm daughters that had been the fodder for a book that would, in less than a year, land in the top 5 of the *Times* bestseller list. In closing that night, Smiley left her Middle American audience with this food for thought: "I invite you to take Kittredge's 'we' personally, for whether we know it or not, as long as we eat, we are involved in agriculture, and through it we are making our world, like Kitteridge's valley, 'a blank perfection of a field.'"

EXACTLY THIRTY YEARS after Meredith Willson's blockbuster film portraying a music man traveling through the small-town Midwest and a woman who falls for his charms, and one year after *A Thousand Acres* took Manhattan by storm, another variant of the woman-on-the-farm theme topped the *New York Times* best-seller list—Robert James Waller's blockbuster *The Bridges of Madison County*.

As with *A Thousand Acres*, the critical reception of the Midwest-set book seemed to hinge on its realism. Unlike earlier farm daughter love stories such as *She Was Only a Farmer's Daughter*, the city man romancing the rural woman in *Bridges* did not come wearing black and sporting a handlebar mustache to mark him as a potential usurper; instead, he arrives on the scene wearing tight-fitting blue jeans. Likewise, *Bridges'* female lead is no Susan Farmer, but a seasoned middle-aged woman who, like Smiley's Ginny, willingly participates in adultery. Even the most casual literary observer could see that by the 1990s farm woman had ceased to be represented as paragons of special virtue; if anything, Smiley and Waller portrayed her, in her lonesome needfulness, as sinful. Less than a month after Oprah's belated visit to the Hawkeye State for what producers dubbed "The Bridge of Love" episode, Waller's one-off novel was named the "all-time word-of-mouth bestseller" by the editors of *Publisher's Weekly*, and industry analysts predicted it would become the all-time leading bestseller for a hardcover novel.

Across the Heartland the debate about the book's representations of Midwest farm wives and daughters heated up in the fall of 1993 with the release of Waller's sequel, *Slow Waltz in Cedar Bend*, and with the location scouting begun by Steven Spielberg, who had bought the movie rights. Charles Champlin, writing in the *Cedar Rapids Gazette*, bristled at the similarities in the two books' heroines. "Waller's first heroine was another man's wife, of extraordinary but stifled sensuality," Champlin noted. "His new heroine, Jellie Braden, is also another man's wife, of extraordinary but stifled sensuality." Husbands, Champlin argued, merited better treatment in Waller's work: "The husband in question is this time more than a stock figure, indeed a sad and subtly, if quickly drawn, personality. Minor characters are a bit more than dim snapshots."

When the location for the Clint Eastwood-directed *The Bridges of Madison County* was announced in 1994, the ironies surrounding Waller's love story began to mount. The Cumming, Iowa, farmhouse with four fireplaces and a big front porch that producers had selected in a fly-over turned out to be owned not by an Iowa farm family, but by an out-of-state absentee landlord. Another difficult irony confronted the big-budget *Bridges* juggernaut when the right to use the location scouts' original, first-choice farmhouse in Winterset, Iowa, was refused on principle by co-owner Aaron Howell, an 84-year-old farmer who had been married to his wife Lola for sixty-three years and "wouldn't stand for any hanky-panky" of the type portrayed in the on-screen romance.

Soon after the story of Howell's refusal hit the newswires, Knight Ridder and other news services descended on Winterset for the opinion of Lola Howell, a farmer's daughter who had lived since 1930 on the same farm and who sided with her husband on the book's tawdry representations of farm women. Francesca's affair with Robert Kinkaid struck Lola Howell as improbable, if for no other reason than, as she put it, "It is not the average Iowa farm wife who stays home when the rest of the family goes to the state fair." When asked whether adultery of the kind indulged in by the book's hero and heroine happened much in these parts, Howell bluntly told reporters, "I never heard of such a thing." Lola Howell hadn't read the book—like her husband, she refused to on principle—but she would no doubt have found the storyline concerning her native state and its farm women all-too cliché.

Admirable as Howell's conscientious objection may have been, nothing could stop the popularity of Waller's fictional wives, who, with the 1993 publication of the sequel novel *Slow Waltz in Cedar Bend*, suddenly stood at number one and two on the best-seller list. In fact, the book that had done so much to raise the ire of and both *Time* and *Newsweek* magazine—who had called its author "Mushmeister" and "The Barnum of Romance" respectively—had done little more than follow the long-established formula for Heartland romances. Its leading man, Robert Kincaid, hails from the state of Washington, but he is, as the recipe dictates, an itinerant at heart, a citizen of the world, described as "all lean muscle, muscle that moved with the kind of intensity and power that comes only to men who work hard and care for themselves." Francesca can't help but fall under his spell when her husband and children leave for the Illinois State Fair to exhibit a prize steer that, Waller sees fit to insert, "received more attention than she did." For Francesca, even Robert Kincaid's diet embodies exoticism. "In contrast with local folks, who fed on gravy and potatoes and red meat, three times a day for some of them," marvels farm wife Francesca, "Robert Kincaid looked as if he ate nothing but fruits and nuts and vegetables. Hard, she thought. He looks hard, physically. She noticed how small his rear was in his tight jeans—she could see the outlines of his billfold in the left pocket and the bandanna in the right one—and how he seemed to move over the ground with unwasted motion."

In the story Francesca is a farmer's wife, an Italian by birth rather than a midwesterner, a detail Waller seems to have concocted to make plausible her simmering passions—passions, we are left to presume, that would not be plausible to readers if sprung from an actual Midwest farmer's daughter. Waller has Francesca wonder aloud, "What was the barrier to freedom that had been erected out here? Not just on their farm, but in rural culture. . . . Why the walls and fences preventing open, natural relationships between men and women? Why the lack of intimacy, the absence of eroticism." While the men farmed, the unfulfilled women of rural Iowa, Francesca muses, "sighed and turned their faces to the wall in the nights of Madison County." Later, when Francesca and the fly-by-night photographer make love while her husband does his best to shepherd the kids at a distant fair, Waller reveals that Francesca has not had an orgasm in years, though her new lover can make her have them "in long sequences" sufficient to help her "become a woman again."

Not long after the sequel to his breakthrough novel Robert Waller took his estimated $11 million in profits and set up shop in Texas, the Heartland perhaps too small to house his fame. "I've moved on from Robert and Francesca," Waller told USA Today's Craig Wilson in 1993. "I have other worlds to go on to." Similarly, circa Smiley's 2008 interview at the University of California, Davis, any pretext that her Pulitzer Prize-winning novel had actually been intended as a redemptive story of empowerment tailor-made to the Midwest had been unceremoniously dropped, as had any laudatory mention of the Heartland as setting.

Entertaining the questions of KXJZ's Jeffrey Callison at a safe remove from the farm daughters whose lives she had represented, Smiley spoke with clinical detachment about the craft choices she had made in the novel that had, in the interim, become the definitive portrayal of the rural Midwest in the eyes of many critics. "What was it about Iowa?" Callison asked the esteemed authoress, eliciting this response: "In order to write the novel, I needed a sort of isolated place for it to take place." Elsewhere in Smiley's comments, the Heartland figured merely as a consideration of craft and coincidence. "One time I was in Upstate New York," Smiley said, revising her explanation of the novel's origins. "It was in a McDonald's actually, and I was looking at pictures of the Iowa landscape. And the person who was sitting with me said, 'You could set that King Lear book in Iowa. And I said 'Bingo.' That is how it came about."

Waller and Smiley hadn't managed to dismantle the mythos of the farmer's daughter as much as they had initiated a swap of one myth for another under the banner of contemporary realism. While they had succeeded in dulling the patina on an American idol, they had arguably tarnished her in the process, knocking her from her pedestal while asserting that she was just as prone to sin as anyone else. *A Thousand Acres*, in particular, pressed that iconoclastic agenda one step further, suggesting that the farmer's daughter—in her isolation, in her subservience, in her toxic milieu—was in fact more prone to sociopathic reactionism and loveless victimization than her urban counterparts.

Waller, meanwhile, had simply picked up where Allen Eppes's *A Maid in Manhattan* left off, reconfirming the long-established literary trope that rhapsodic love could for the farmer's daughter only be found in the arms of an outsider. Thus, Waller and Eppes made for themselves a double bind—needing to create believable Middle American farm daughters who readers would empathize with when, not if, they shed their homely, native identities for more passionate amours. In each case the authorial answer was to draft the "farmer's daughter" with an asterisk: Waller lent his college-educated Francesca an Italian rather than Iowan birthright while Eppes saw to it that Susan Farmer was a farmer's daughter in name only, making her, in reality, a high-bred, well-meaning socialite condemned by circumstance to an orphan's life on the farm but determined to repay her Aunt Alice—the genuine article—for her benevolence by posing as a true-blue farmer's daughter. In the end Eppes, Smiley, and Waller helped make Midwest farm women romantic again, but at a considerable cultural cost.

PART TWO

"Springtime in the country. City children may well envy their little country cousins the free life in the open and the companionship with animals."

CHAPTER EIGHT

LITTLE HOUSES
ON THE PRAIRIE

"**T**RADING TIARAS FOR Calico on the Prairie," ran the headline in an August 2011 *New York Times* dispatch from De Smet, South Dakota. Reporter Anna Bahney had what surely felt like a scoop—a Laura Ingalls Wilder-inspired renaissance in a midwestern prairie town where, from 2009 to 2010, visitor spending had jumped 27 percent, and in July 2011, county revenue from a tourism tax was up a whopping 22 percent, presumably on account of Laura.

"Many parents are finding that Laura and her sisters make welcome alternatives to the Disney princesses Belle and Mulan," the *Times's* Bahney observed. Cheryl Palmlund, executive director of the Laura Ingalls Wilder Memorial Society in De Smet, added more anecdotal evidence to the heft of twenty-thousand-per-year visitor numbers. "There are more little girls coming dressed and ready to be Laura," Palmlund said. "They come out here to play on the prairies and they know all the stories."

Of late Wilder enthusiasts have made big business of surmising how the world's most famous pioneer girl would react to any given contemporary bugaboo. WWLD— What Would Laura Do—seems an ironically pitch-perfect acronym for a distracted digital age. In spirit, it's everywhere, from Dr. Timothy Walch's introduction to a recent Wilder Christmas exhibit at the Hoover Presidential Library in the *West Branch (IA) Times*, in which he observed, "Laura Ingalls Wilder can be a tonic for troubled times," to children's writer and suburban goatherd Deanna Caswell's commentary for the *Memphis Commercial-Appeal* entitled "Practically Green: To Cut Back on Waste, Ask 'What the Wilders Would Do?'" Caswell shared this contemporary chestnut in her op-ed: "The children and I listen to audio books in the car, and this month it's *Farmer Boy* by Laura Ingalls Wilder. I'm absolutely blown away by their Greenness. The theme that keeps repeating over and over is 'waste is sinful.' When Almanzo feeds candy to the pig, he worries that he will get in trouble for 'wasting candy on a pig.' And when he's carrying eggnog to the field workers, he is careful not to spill because 'waste is sinful.'"

Meanwhile, author and erstwhile midwestern farm girl Jennifer Worick made WWLD the thematic underpinning *The Prairie Girl's Guide to Life.* "Perhaps you'll be inspired to sit down—along with a perfect pot of tea and thick slice of rhubarb pie—and talk for a spell about your kin," Worick invites her female readers. "They might share some memories with you and, in return, maybe you can teach them a thing or two. Come to think of it, it sounds like just the kind of rewarding thing and industrious visit Laura would have enjoyed."

Murkier is what, exactly, a new generation of Wilder fans visiting childhood sites in quintessential midwestern towns like Pepin, Wisconsin; Independence, Kansas; Burr Oak, Iowa; Mansfield, Missouri; and De Smet, South Dakota, hope to find on the prairie, if not the simple reassurance that the life of Wilder is still possible for their children, or their children's children. Wilder herself was known to question the utility of such strolls down memory lane. In a 1946 letter to midwestern schoolteacher Ida Carson, the author recalled several trips she had taken back to her childhood stomping grounds in De Smet, where, she noted, she came away "unsatisfied." The old homestead and its homey places, she lamented, had so changed from "the old, free days that we seem not able to find there what we are looking for. Perhaps it is our lost youth we were seeking in the place where it used to be."

Wilder's nostalgia for the leaner and harder years of her youth begs a larger question: how many of today's De Smet pilgrims would willingly choose for their daughters or sons the character-building hardships of farm daughter Laura's life? More to the point, how many have put a down payment on that existence, voting with their feet for a life on a working farm? Perhaps acolytes come by the carload seeking not a real-life return to the yeoman's life, but a symbolic rekindling of Wilder's iconic farm girl values—patriotism, teamwork, freedom to live free of interference. If so, why choose a century-old pioneer girl to carry the weight of a contemporary longing? Is it because Wilder, among all of literature's admirably enterprising and plucky young women, is most emphatically rural that former first lady and school librarian Barbara Bush lists Wilder's books as among her favorites? Is it because little Laura represents the kind of agrarian for which Jefferson and other founders longed that MacArthur once arranged for the translation of Wilder's work into German and Japanese as part of the State Department's Americanization efforts?

GET A GROUP of generation X farm kids together, add merlot, and chances are better than good they'll start reminiscing about *Little House on the Prairie*. Do the same with a mixed group from my own generation in a cabin that sleeps seven, turn out the light, and wait for the first ham to call out into the darkness, "Good night, John-Boy. Good night, Mary Ellen."

For a generation who grew up in TVland, screen visions of the farm—and farm girls—often supplanted the real thing, even among the real thing. The situation becomes, as graduate students like to say, very "meta-meta": the farm daughter who grew up in an age of large-scale farming when, often as not, she had few chores to do and few life and death responsibilities to call her own, sees herself reflected in the freckled face of Melissa Gilbert. Gilbert, in turn, impersonates a young Laura Ingalls Wilder, who herself fictionalized her real-life prairie girl experience in her many popular novels.

Likewise, when viewing *The Waltons*, a girl born into the average nuclear farm family of the 1970s can only imagine what it would be like to endure the deprivations of the Depression, or to share a room with six siblings and a house with grandparents—all rites of passage associated with Midwest farmers' daughters, but memes that, by the 1970s, had long ceased to be a reality for most farm girls. Wilder scholars like Anita Clair Fellman and Diane Lanctot see the cultural impact of the books in a postfeminist era as "increasingly complex and volatile," and their heroine as a "feminist girl heroine role model," though Fellman purposefully problematizes such reflexivity. "They [female readers] are often compelled to try to read themselves into texts in which they are either absent or demeaned," she reminds.

Set in pioneer days on the American frontier, the popularity of *Little House*, even more than the *Waltons*, created a strange, postmodern stew for generation X rural kids and their parents to digest. While the real-life Wilder lived that of which she wrote, her pioneer girl accounts had been thoroughly fictionalized even before executive producer Michael Landon turned them into the stuff of network television. And yet for flesh and blood farm girls born of the 1970s, *Little House's* set—meant to simulate Walnut Grove, Minnesota, in 1932—was often as close as they would get to a life hand-hewn from the land—a life without air-conditioning, a life where everyone had to buck up and complete life-or-death chores. Ma and Pa, Lindsay Greenbush, and Michael Landon, were as real, and sometimes more real, in generation X imaginations, than real-life farm kin who were typically busy in the 1970s either growing family farm operations to factory size or turning to off-the-farm work to make ends meet.

Statistical measures signaled tough times ahead for Midwest farmers' daughters like my sister, who came of age in the 1980s. In 1920, when Wilder was busy penning her pro-rural commentaries for the *Missouri Ruralist* and *Star Farmer*, and three years after my great-grandfather first plowed his wife's land, the farm share of the total US population stood at just under 30 percent. By 1950, three years after he dedicated his one and only book, *The Furrow and Us*, to his own Midwest farmer's daughter, Helen, the percentage hovered around 15. In 1980, when my sister entered fourth grade, it stood at around 3 percent. In 1950, when my mother was born on farm in Linn County, Iowa, the USDA's own maps classed the vast majority of the counties west of the Mississippi River and east of the Sierra Mountains as "nonmetropolitan" farm communities, including nearly three quarters of my home state of Iowa. By 2000 the USDA Economic Research Service could find only a handful of truly farm-dependent counties in a Hawkeye State then fully 65 percent urban.

To Midwest kids internalizing such profound geo-cultural shifts, Ma and Pa Ingalls sometimes seemed like the parents they might have wished for rather than the Nixon-era ones they had, and especially so for generation X, for whom TV often played both teacher and jinni. My friend Becky Kreutner recalled of that time: "My main [female] role models were definitely within my family, especially my mom and my grandma. At first I couldn't think of any others, but I would have to say that Laura Ingalls Wilder. . . . When I was five or six, the show *Little House on the Prairie* first aired on KWWL. After the pilot, I wrote a letter to the TV station telling them how much I enjoyed the show."

While the show and its actors and executives were roundly panned by critics, conservative groups and moralists loved the TV adaptation of the classic series, and a generation of Midwest farm girls grew up dreaming that the image of Melissa Gilbert and Michael Landon on their television sets was the purest representation of their heritage, a clear antecedent to their own agrarian resourcefulness. For her part Wilder would likely have quibbled with the show's portrayals of her family and friends, but not with its championing of pioneering values, for Wilder herself believed raising children in the country to be superior. In a February 1911 column for the *Ruralist* she wrote:

> Yes indeed, things have changed in the country, and we have the advantages of city life if we care to take them. Besides, we have what it is impossible for the woman in the city to have. We have a whole five acres for our backyard and all outdoors for our conservatory, filled not only with beautiful flowers, but with grand old trees as well, with running water and beautiful birds, with sunshine and fresh air, and all wild, free, beautiful things.

The children, instead of playing with other children in some street or alley, can go make friends with the birds on their nests in the bushes, as my little girl used to, until the birds are so tame they will not fly at their approach. They can gather berries in the garden and nuts in the woods and grow strong and healthy, with rosy cheeks and bright eyes. This little farm home is a delightful place for friends to come for afternoon tea under the trees. There is room for a tennis court for the young people. There are skating parties in the winter; and the sewing and reading clubs of the nearby towns, as well as the neighbor women, are always anxious for an invitation to hold their meetings there.

In conclusion, I must say if there are any country women who are wasting their time envying their sisters in the city—don't do it.

Elsewhere in columns with titles like "Learning to Work Together," Wilder relayed parables of farm upbringings as she witnessed them on her 5-acre Ozark farmette. Always, their effect is to demonstrate that allegedly urban values such as selfishness and competition must not be allowed to reach farm daughters like her own Rose Lane. "Cooperation, helpfulness, and fair dealing," she wrote in 1916, "are so badly needed in the world, and if they are not learned as children at home it is difficult for grownups to have a working knowledge of them." If country-dwelling mothers failed to realize their influence on the next generation, Wilder was there to remind them, "The hand that rocks the cradle is the hand that rules the world."

By the 1970s and early 1980s, when a generation of Midwest farm girls like Kreutner began to watch *Little House on the Prairie*, readership of the by-then fifty-year-old books had begun to flag, and a whole new medium was needed to make the book's original nostalgia new again. Conservatives at the time pointed to such calamites as Watergate, rising divorce rates, and a string of military failures overseas as prepping the nation's families for a nostalgia cure, but the same rhetoric had, in fact, been applied to Wilder's *Little House* books on first appearance. Heralding the series in 1941, Annie T. Eaton of the *New York Times Book Review* observed that "the history of our country and the doings of our pioneer forebears have a special significance for us." Two years later writer Irene Smith wrote that *Little House* served as a necessary reminder of the "canniness of the pioneer, the strength and joy of the builder."

The same proved true thirty years later, when, among a generation of farm families, the *Little House* gang offered the timeless balm of a pioneering farmer's daughter. Despite sentiments like those of *Chicago Tribune* TV critic Gary Deeb, who called the series debut a "meatless sausage of cloying sweetness, padded dialogue, and soap-opera background music, all brought together by a lisping little girl narrator," Middle Americans loved it when Melissa Gilbert said things like "home is the nicest word there is." A few short years after the series premier, the National PTA issued its report card, and *Little House on the Prairie* was selected the top program.

Early on, news that NBC's successful two-hour debut of *Little House on the Prairie* would be turned into a regular series was greeted with celebrations by 1970s back-to-the-landers, including Roger L. MacBride, the grandson, by adoption, of Wilder, and himself what a 1974 AP report called "a gentleman farmer in Charlottesville, Virginia." MacBride reported that he was "beaming all over" at the news that his long-dead grandmother would land her own TV series. He had been up all night, he said, when the AP's Joy Sbarbutt called for comment. "All of a sudden, about 2½ years ago, there was a mob descending—through the publisher, on me, wanting to buy the rights," MacBride enthused.

But what had begun as a promising relationship between Wilder's heir and NBC attorneys turned, in the space of a few years, into a nightmare. By 1978 the books had indeed been transformed into a series, NBC's highest rated, despite widespread criticism of star and executive producer Michael Landon having what Norma Lee Brown termed an "Orson Welles complex." Brown quoted Ed Friendly, creator of *Laugh-In* and a creator-producer of *Little House*, as saying, "Even if the series is a TV success, which it may be, it will not be one of which I am proud." Ditto for MacBride, who, after working on the pilot with Landon, jumped ship, saying, "In my opinion it is a weekly package of garbage unrelated to her books," he told Browning, adding that the problem was Landon. "It's all Hollywood made-up stuff, much of it an outrageous copy of *The Waltons*. It has no fundamental guts to allow it to last."

But last it did, piling up Emmys and turning Middle American moms into Landon lovers despite what negative publicity he received for being the show's dictator. In many a mom's eyes Landon was a dreamboat, the "Big Hunk on the Prairie," as *Little House* star Alison Arngrim put it in her memoir, *Confessions of a Prairie Bitch*, describing the titillation inherent in seeing Pa Ingalls "stripped to the waist, glistening with sweat, grabbing his wife around the waist with a lust not normally publicly displayed in the 1800s." In 1975, even as Friendly and MacBride went on a media offensive, Landon's popularity among the general public continued to grow. Profiling the star in the *Chicago Tribune*, Anita Gold wrote, "And pssssssst . . . if you want to know a secret about something Landon does, well—he writes his wife love letters once a week and sends her roses. As far as I'm concerned the only awful thing about Landon, is that he's awful nice."

If anything, the popularity of *Little House* among rural families showed their willingness, if not eagerness, to accept a fantasy projection of pioneer life. Critics such as Norman Mark of the *Chicago Daily News* drew attention to the series' almost delusional effect, calling the show a "big disappointment" and a "perversion." Mark and others pointed out the many discrepancies between the novels and the show—Laura's windows were too wide; the Ingalls were too close to town; the weirdly clean-shaven Pa was too ready to deal in cash. "But the worse transgression against the Wilder book is on the storylines," Mark lamented. "In the book the howling of the wolf, the passing of an Indian tribe, the scream of a mountain lion, the arrival of a window, the digging of a well, the singing of Pa's violin . . . and the arrival of Christmas were memorable family experiences."

Midwesterners loved Landon, however, not just for bringing a family story of a farmer's daughter to television at a time when nearly all other rural shows had been axed, but because he seemed to share an innate Middle American distrust of Hollywood. Commentaries like Jerry Buck's for *Tempo TV* relayed stories of how closely the family-oriented Landon scrutinized his own children's TV-viewing habits, quoting the megastar's admission, "We don't let our children watch television on school nights unless my wife and I judge it will have some value for them. Then we talk about what we've seen." Buck's article, "Landon Proves his 'House' Not Made of Cards" ran in the same year, 1978, that *Little House* scored a tie with *60 Minutes* in the Nielsen Ratings.

But if the goody-two-shoes prairie routine Landon proffered seemed too good to be true, too much "sweetness and light" as John O'Connor of the *New York Times* complained in 1974, the comeuppance came in 1980, when word first reached the cast and crew that Landon's storybook marriage to wife Lynn was ending. In her memoir *Prairie Tale*, Melissa Gilbert wrote of the seismic event: "In public Mike was seen as a pillar of morality and family values, a real-life incarnation of Charles Ingalls, not someone who would leave his wife for a younger woman." Gilbert noted that everyone had bought into a myth, creating a long way for a TV actor playing an irreproachable pioneer dad to fall. "The shit hit the fan in the tabloids and celebrity magazines. Mike lost lucrative commercial endorsements. He admitted he wasn't perfect and warned people not to confuse him with the character he played on TV," Gilbert recalled.

Back in eastern Iowa, where my almost-teen sister was just then weaning herself from *Little House*, a string of *National Enquirers* bearing Landon's mug arrived at our rural route. Still, we kept on tuning in, something about the show itself and its impact on farm families bigger than one man's peccadilloes. When the series finally did appear to be coming to the end of its life cycle in the spring of 1982, NBC big fish and chief executive Grant Tinker called on the beleaguered Landon personally in an attempt to secure his return to the show. A whole generation had grown up with the series, then in its eighth year, and the thought of a Monday night without *Little House* felt like a loss of innocence—almost like leaving the farm itself.

Though the series ran for another year without Landon under the title *Little House: A New Beginning*, endings proved hard to stomach, not just for audiences, but for cast members who had grown from children to young adults on a set depicting TV's version of Walnut Grove, Minnesota. Arngrim recalled in her memoir, "It was a tough time for all us girls on the prairie. We were no longer children, we were young women, but people had a hard time visualizing us that way. We all felt this enormous pressure to 'break out' of our 'wholesome' image. Melissa Sue Anderson was playing an axe murderess in the lurid 1981 slasher *Happy Birthday to Me*, Melissa Gilbert was wearing twelve pounds of eyeliner and running all over town with Rob Lowe on her arm (among other body parts), and I was popping up in the *National Enquirer* every other week in a bikini."

The fraught denouement of *Little House* symbolically paralleled the dilemma faced by real-life farmers' daughters like my mom and sister, who wondered at times, as every farm girl must, when, if ever, does one stop being a farmer's daughter? When does the forever-daughter, typed if not typecast, find herself cast in larger roles? Is one always a farmer's daughter, like one is always a senator? *Little House*, after all, had turned into a veritable spirit-possession for nearly everyone associated with it—for Landon, who could never quite reconcile the image of Pa with his own far-from-perfect moral life, for Gilbert, who recalled that once she slipped on the dress of Laura Ingalls Wilder she "felt different, transformed into someone else, as well as transported back in time."

Gilbert, sixteen when Laura's wedding with Almanzo aired, was, in her personal conservatism, indicative of the conservative values of the show's audience, who both wanted Laura to grow up, and continued to tune in because she didn't. "My head was full of vivid images of the vile things that could happen to me in there," Gilbert recollected of her first bedroom scene with Dean Butler. "Not that I had a clue what any of those things looked like; for me it was all imagination, and what I was imagining scared the shit out of me." The show's viewers, most of whom had lived through the 1960s, knew exactly what was happening under the prairie sheets between Laura and her new husband, but somehow willed themselves to enlightened innocence. Gilbert remembered in her memoir that, for the sake of realism, she had had to wear a padded bra in *Little House* because, "once you were a woman on that show, you couldn't be smaller than a B cup."

In September of 1983 the *Burlington (IA) Hawk Eye* ran vivid news of the show's demise that farm people especially could understand—the buildings had to come down. In a plot device contrived to return the leased land on which a simulated Walnut Grove had been erected, Landon announced he would return for a final, two-hour series finale, slated to run in 1984 under the apropos title, "The Last Farewell," and scheduled to show each cast member pushing plungers to dynamite their own homes. The reason for the destruction—unwanted Easterners moving in—also struck a Middle American chord. NBC publicist Bill Kiley described the final episode as a classic frontier confrontation where "a group of eastern financiers, the bad guys, reveal to the townsfolk that they own the property, and . . . that from now on they are their employees." Though the plot might have come straight from central casting, midwesterners could relate all too well, as a substantial number of small farmers had by the early 1980s begun to lose their land to large, out-of-state syndicates who sometimes rented it back to its one-time owners, in many cases at a steep markup.

Our collective grief was partially assuaged when we learned *Little House* would live on in reruns in select markets like ours. Every day well into the late 1980s, by which time I was old enough to drive and my farm daughter sister was already off to college, *Little House* would grace the television set each afternoon after *The Brady Bunch*. Meanwhile, my older, farm-reared cousins had already begun to push the plunger on their childhood selves, blowing up once-rural identities in favor of newer, savvier calling cards as college students in cities like Chicago, Iowa City, and Kirksville, Missouri. Among the cousins, all the farmers' daughters had gone away to the city, the only grandchild left "on the set" of our own little farm on the prairie was me. "Nothing lasts forever," my uncle Paul, an old farmer, always said—not even childhood.

CHAPTER NINE

FUTURE FARM DAUGHTERS OF AMERICA

"**HAVING A CHANCE** to meet with some of the men and women in this room has only made me feel more confident," President Barack Obama declared from the makeshift lectern at the front of the room—a steel-sided pole building at Northeast Iowa Community College in Peosta, Iowa. The topic for the day was rural economics, and the President and his staff had invited approximately one hundred ag stakeholders and ag writers and journalists to join the conversation. Prominent among us were the blue-jacketed members of the Missouri, Illinois, and Iowa chapters of an organization now known simply by its acronym FFA, having officially dropped the moniker Future Farmers of America.

This is not your father's FFA. Founded in Kansas City in 1928 by a group of thirty-three midwestern farm kids on hand for a livestock convention, more than half of FFA's state leadership positions are now held by young women, including Lindsay Calvert, the newly elected president of the Iowa FFA and an ag communications major from Guthrie Center. At the President's forum Calvert had frequently been culled from the crowd to pose for network cameramen perched atop media risers. Each time she smiled beatifically for the lens in her trademark navy blue jacket, obligingly turning so the camera could zoom in on the dinner plate-sized insignia worn on her back like a bull's-eye.

"I saw some of these Future Farmers of America and their young president right over there," Obama had said, gesturing in Calvert's direction, "and when you hear the enthusiasm and energy that these young people display, and the fact that if they can just get a little bit of a break when it comes to getting started on the front end, get a little bit of help with capital, they are ready to take American agriculture to the next level—it gives you confidence, it gives you hope."

FARM YOUTH, AND especially farm daughters, have always been turned to in periods of economic and cultural uncertainty. "The seed of the 4-H idea of practical and 'hands-on' learning came from the desire to make public school education more connected to country life," the 4-H National Headquarters history page reads. "During this time researchers at experiment stations of the land-grant college system and USDA saw that adults in the farming community did not really accept new agricultural discoveries. But educators found that youth would 'experiment' with these new ideas, then share their experiences with the adults."

In a sense 4-H and FFA have historically served as youth focus groups for revolutionary ideas not just about crops, but about the *culture* in *agriculture*—gender roles figuring prominently in those formulations. Young rural women especially, considered in the early days of 4-H to be more inherently social than their male counterparts, could be highly effective disseminators of official messages, if rural society could first be reformed to include opportunities for systematic socialization. Toward that end and others Congress passed the Smith-Lever Act of 1914, which created an extension service to be housed at the nation's land-grant universities and charged it with keeping the citizenry informed regarding research and best practices in agriculture and home economics.

The Smith-Lever Act, writes T. T. Martin in his 1956 monograph, *The 4-H Club Leader's Handbook*, provided a "legal basis" for the 4-H club work that would follow. While 4-H operated as an out-of-school program for youth as part of the Extension Service, FFA operated as part of the regular school program under the auspices of the Smith-Hughes Act. The two sister organizations were forever being confused, lamented Kenneth B. Hoyt, director of career education in the US Office of Education, in his monograph, *Future Farmers of America and Career Education*, but the basic differences were simple: FFA concentrated on cultivating leadership and citizenship skills among students ages 14 to 21, while 4-H focused on the "total human development" of young adults and children. In either case both government-supported organizations flourished in the decades between World War I and II, powered in part by a renewed interest in food production.

The first of the Midwest farmers' daughters in my clan to join 4-H were my Great-Aunt Mary Puffer and her sister Julia. Throughout the 1920s the *Pioneer Press* in Mechanicsville, Iowa, reported the progress of the new "Pioneer Peps" 4-H Club that great-aunt Mary helped found. "The Pioneer Township 4-H club met at the home of Mary Puffer Thursday, May 30. Election of officers was held," the *Pioneer Press's* first bulletin read. "Officers elected were: President, Julia Puffer; Vice President, Jean Thompson; Secretary and Treasurer, Lucille Cook; Historian, Ada Mae Harmon; Publicity Chairman, Eleanor Joy Nicoll. Several songs for rally day were practiced and also the different tests of flour were made." A follow-up blurb contributed by Club Reporter Joy Nicoll read, "The Pioneer Peps 4-H club met June 8 at Mary Puffer's for an all-day meeting. The lesson was on whole wheat bread. Each girl kneaded and shaped her own sample loaf like the bakers shape their loaves. Sweetbreads, four-leaf clover rolls, twisted rolls, triangular rolls and kolaches were made. They also made an apple coffee cake, an apple crescent and a Swedish tea ring. They had three new members, Eva Aaron, Barbara Ferguson and Ruth Ferguson. The girls will give a demonstration at the Farm Bureau meeting Friday evening. Songs for Rally Day were practiced with the Sunshine Circle Club Monday. The two clubs will present 'The Romance of Bread' for Rally Day at Tipton, June 13."

A final newspaper item detailing the Club's inaugural year reports a third in-home meeting. "Pioneer Peps 4-H Club girls met at the home of Julia Puffer Thursday, August 1. The demonstration team, Julia Puffer and Jean Thomson, gave a demonstration 'A Quick Bread for Various Occasions,' and baked a loaf of honey nut bread. Singing and planning for the county fair exhibit came after. The next meeting will be at the home of Marion Brown either August 15 or 22. All girls will help give lessons on yeast, cheese, table manners, and posture."

It's hard to imagine my free-spirited and eccentric grandmother, the young Julia Puffer, as a "table manners" kind of girl, but that she presided over such lessons suggests the 4-H message—to develop the head, hearts, hands, and health of the most promising young rural girls and boys—was being heard. Indeed, to be a farmer's daughter from a "civilized" farm family in the 1920s meant having the ear and the interest of national planners intent on recruiting the country's most talented sisters to serve as boosters for small-town and country life. Almost from the get-go 4-H had the Puffer girls in its sights.

On Julia's wedding day on December 11, 1941, the United States declared war on Italy and Germany, and the perennially isolationist Corn Belt found itself suddenly recast as interventionist. A front page story in the *Cedar Rapids Tribune* that morning trumpeted, "Iowa, conservative Midwestern state which last week would have voted down any proposal to enter upon a foreign war—Iowa has changed its mind on that point. Not in any of the forty-eight states of the Union is patriotism running higher or support of Congress and the President more wholehearted than in Iowa today." The US Naval recruiting station at the post office, the article reported, saw an estimated five hundred men attempt to enlist in less than forty-eight hours. For the children of farmers, many of whom planners believed were needed more on the home front than on the ground overseas, 4-H offered parents and their children the appealing prospect of what brochures termed "normal activities for young people in abnormal times," helping to staunch "the increase in juvenile delinquency which comes from war periods."

One such pamphlet, produced by the University of Illinois College of Agriculture and the USDA, entitled titled "Boys and Girls Can Help," bore the fine print: "This folder is for FATHERS AND MOTHERS and others interested in Rural Boys and Girls." As before, the language of the 4-H belied more complex societal motives. Enrolling one's children in 4-H in the wartime Midwest was presented as a patriotic duty. "Approximately 32,000 Illinois boys and girls enrolled in 4-H projects in 1942," the literature claimed. "With the help of fathers and mothers other interested people, this number can be doubled in 1943." The brochure exhorted rural parents to "give boys and girls a place in the war effort." The 4-H youth army, the USDA realized, could be a valuable ally in the battle for food supremacy, thus the literature made a point of citing the 3 million bushels of vegetables, 6.5 million chickens, 300,000 hogs, 85,000 dairy cows, and 14 million jars of food organization boys and girls had produced in 1942—enough to supply an army of 150,000 men for a whole year, the organization pledged.

The 1950s *4-H Club Leader's Handbook* makes the socialization agenda of the program still more clear, stating "incentives in club work have been set up to appeal to youth so as to recognize and meet their apparent economic, social, and educational needs, such as the following: to produce food for people and feed for livestock; to increase farm income, the ownership of property, and family partnerships; to develop vocational skills and democratic group action; to round out character through citizenship responsibilities; and to engender an appreciation for the finer things of life. All of these purposes are being achieved without the exploitation of youth."

While the *Handbook* conceded that the club had been "closely associated to living on a farm," Martin anticipated the need for his organization to both recruit in, and serve, urban America, a move that postwar demographic trends indicated would be necessary. He suggested remaining "elastic" as the key to survival for 4-H club leaders during a time of rapid suburbanization. Most of all, the author noted, aggressive, targeted recruitment must continue if 4-H hoped to grow its ranks, and should be concentrated, as he put it, "on 10-year-olds, rather than on older boys and girls, who are poorer enrollment risks. Recruitment should not stop with the child," Martin emphasized, but should extend to the parents and "eligible brothers and sisters and best friends."

In 1952 one of those 10-year-olds was Julia and Edward Jack's first child, Patricia, who had grown into a budding seamstress not so much with her mother's help, but via the encouragement of an army of 4-H leaders who had stepped in to help America's mothers raise a generation more than willing to pledge their head to clearer thinking, their heart to greater loyalty, their hands to larger service, and their health for better living for club, community, and country. "4-H was really important to me," Patricia recalls. "It was a social and learning thing. I learned to do alterations and to do so many other things as simple as how to make a bed. My mom didn't have much time to teach me with a younger family to take care of. As far as I remember, only a very few [4-H leaders] were farm wives. Being a farm wife was a full-time job."

Characteristic of her generation, my maternal grandmother had become more of a specialist than her mother before her, my mom claims, adding, "She didn't have outside duties. They separated the women's work from the men's work back then. . . . She didn't even know how to drive a tractor." At the same time, with five children underfoot, my mom's mother didn't have time to share with her girls the finer points of the domestic arts she had come to perfect. "Mom did not teach us about cooking . . . which I thought was pretty unusual," mom recalls. "Interestingly enough, when it came to the kitchen, we really weren't taught anything."

In the eyes of many daughters of the time, farm wives of the 1950s and early 1960s looked increasingly like second-class citizens. Mothers were busy, heroically so, but not necessarily in the outdoorsy, hands-on way of the 1920s rural women of whom farm scholar Mary Meek Atkeson once wrote, "The woman on the farm . . . is perhaps the busiest woman in the nation" because she was the "arbiter of health and happiness." Based on the countless letters penned her by Roaring Twenties farm women, Atkeson felt confident in declaring that the woman on the farm was "an important factor in her husband's success" especially if she also maintained a side career and took part "in a wide variety of community affairs." "If she is the busiest woman," Atkeson summarized, "she is also one of the happiest. She finds little to complain of in her lot."

But for my mom, and for many of her generation, farm mothers did not seem especially happy, nor especially instrumental in the nuts-and-bolts operating of the farm as a business or as a springboard to wider community outreach. "A farmer who did well and made money and had good crops and nice fields was really looked upon as a leader in our community," mom recalls. "My dad and my mom didn't take on leadership roles in anything but the farm, except maybe church. They were never interested in politics, either local or national. As I got older that was always an embarrassment to me . . . that they didn't vote."

The midwestern 4-H farm girl ensconced in an age as conspicuously politicized as the 1960s often found herself in gendered limbo, caught in a generational divide, increasingly separate from her mother and her mother's plight, but not yet having been granted access to her father's world. My aunt Patricia, for example, doesn't recall ever quibbling over the sexist 4-H projects of the mid- and late 1950s, though perhaps that's because her interests—art and home economics—already aligned closely with approved subjects for girls. 4-H home economics projects for Patricia and her 14- to 20-year-old peers included, for example, in the food category, "Plan and Prepare Meals (Pastry)," "Yeast Bread in Your Meals," "Canning," "Freezing," and "Outdoor Meals," among others, while under the category "Room Improvement" fell such opportunities to shine as "Flower Arrangement," "Pictures for Pleasure," "Table Covers for Dining," and "The Five-Year Room Plan." Projects open to boys or girls jointly included two options—the self-explanatory hosting entailed in "Party-A-Month," or an exploration of "Handicraft," whose vaguely worded minimum requirement read "make 5 articles." Meanwhile, projects open to rural sons included more physically ambitious tasks such as "Metal Roofing," "Tractor Care," and "Farm Concrete."

1956, when Patricia turned 14, marked the release of the first mass-market book of its kind concerning 4-H, *The 4-H Club Leader's Handbook*. The book's author, T. T. Martin, had worked as a club agent in Missouri. In the book's foreword Missouri's Charles E. Lively opined, "This book is long overdue, and yet such a book could not have appeared until a person with a thorough knowledge of 4-H club work and considerable familiarity with the social sciences . . . was willing to write it. That man has proved to be T. T. Martin, for many years a 4-H club supervisor." The 4-H club, Lively claimed, had proven to be "one of the most successful phases of the Agricultural Extension Movement . . . during the past 40 years." It was to be commended, he wrote, as a "system of practical education using instructional materials indigenous to the environment in which rural youth live."

Chapter two of Martin's *Handbook* cut to the chase, describing for would-be leaders need-to-know similarities and differences between farm boys and girls. While boys and girls still pursued separate projects, the sexes sometimes mingled in meetings, and this, Martin believed, meant that leaders needed to be advised of basic physiological facts—that girls reached puberty an average of two years before boys and that boys were significantly stronger. Normal size, strength, and development made socializing among peers easier for farm youth whose characteristics fell within norms, Martin cautioned, and became subsequently more difficult for those young adults exhibiting a "variance from the normal" that might cause them to be viewed as "less acceptable age mates" and thus potentially suffer a loss of "prestige." The datedness of certain aspects of 4-H experience in the 1950s manifest clearly in the handbook, wherein the author voiced such politically incorrect notions as: "Fat boys and girls are often unhappy and may be maladjusted because of glandular disturbance. Girls of this type are often known as 'tomboys.' Later, older girls and women who are overweight and do not consider themselves attractive may assume masculine roles in order to gain distinction." However, Martin emphasized, "There is no intellectual superiority in either sex. The difference lies only in special interests and abilities, girls usually excelling in language and boys in mechanics."

By the time the British Invasion landed, sending my aunt Barbara and my father, Michael, scurrying to the Hardacre Theater in 1964 to see *A Hard Day's Night*, their eldest sister, Patricia, had embarked on an education at the nearby University of Iowa, where she would major in the 4-H-friendly disciplines of art and home economics. Back home, though, Patricia's younger sisters were finding 4-H a bitter pill to swallow. "It was what country girls did," my aunt Barbara recollects. "Patricia thrived in it . . . Susan only wanted to meet 4-H boys, and I just *hated* it!" While the *Pioneer Press* of May 31, 1962, trumpeted news of a 4-H rally that day in nearby West Branch, expected to draw 270 girls from 19 Cedar County 4-H chapters for a day of song and initiation, Susan Jack would have to be dragged along, preferring instead to put her efforts into more rewarding high school activities. Times had changed; whereas her mother and aunt had led the 4-H Club in their day, Susan, the *Pioneer Press* reported, had been elected to the sexier role of class cheerleader for the 1962–1963 year.

Barb and Susan Jack weren't the only ones with an adolescent bone to pick with 4-H. As the Sexual Revolution took hold, Midwest farmers' daughters grew impatient with the limited palette of projects available to them, and many longed to cut their teeth on the more rigorous tasks reserved for the boys. My mom recalls, "It wasn't the kind of 4-H I would have loved, which was animal 4-H. It was the girly stuff . . . the sewing and the baking. I particularly hated it because we had to keep record books, and I've never been very organized. I hated it, I absolutely hated it, but we had to go anyway."

Increasingly, 4-H projects struck many coming of age in the Vietnam era as backward and anachronistic, as seemingly passé as the family farm. While, for example, the headlines in July of 1965 included Lyndon B. Johnson's doubling of the monthly draft from 17,000 to 35,000 men, and Mariner 4's return with the first-ever images from Mars, 4-H leaders like Martin still recommended as a suitable project the decidedly non-space-age "Establishing of a Live Multiflora Rose Fence," expecting enthusiasm of Sixties children upon learning that "a living fence of multiflora rose provides an economical livestock barrier."

By the late 1970s, when my neighbor, Midwest farm daughter Angela Crock, considered 4-H, she found herself turned off by its enduring gender norms. In 1978 Midwest farm daughters were still largely confined to sexist projects such as "Artistic Arrangement" and "Landscaping." During that same year demonstrations from Illinois 4-H members at nationals were presented under seemingly nebulous titles ranging from "Autumn Imagination" to "Cactus Live Dish" to "Asparagus Quiche" to" Making and Selling Terrariums." "I went to one meeting and decided I didn't want to participate," Crock recollects, "because at that time the boys and girls had separate clubs, and, being the outdoor girl that I was, I had no interest in doing the sewing and home projects that the girls' club stressed."

Growing up in nearby Muscatine County was Daleta Christensen, who would go on to become a 4-H extension leader for my home county of Cedar. "I was a 4-Her in the 1970s in Muscatine County," she recalls. "At that time there were not coed clubs, and they referred to my club as a 'townie' club. . . . There were only three project areas offered, and they were rotated through the years so that one year everyone in the club worked on food/

cooking nutrition, the next year would be sewing/clothing and the next year home improvement." Christensen's best memories didn't come from the limited palette of gendered projects, but from what she characterizes as "the fun I had with my friends (softball tournaments, lock-ins, dances) and the support and encouragement provided by 4-H leaders."

Farming itself had suffered something of a public relations defeat felt deeply in the ranks of midwestern 4-H. In 1972 the wonder pesticide DDT had been banned by the EPA, in part as a response to farm daughter Rachel Carson's 1962 exposé *Silent Spring*. In 1971 the *Oelwein Daily Register* in Oelwein, Iowa, ran with the headline "Farming 3rd Most Perilous Occupation," and quoted Secretary of Agriculture and Purdue University graduate Clifford Hardin's contention that the lethal profession, though at that time it employed only 4.5 percent of the nation's workforce, accounted for nearly 20 percent of its work-related deaths. Fifteen years earlier extension safety specialist at Iowa State University, Norval Wardle, reported an estimated one thousand farm fatalities annually were caused by tractor accidents, a number that had nearly doubled in the preceding five years. Farming had always been a dangerous profession, ag defenders noted. Still, half-century-old "expectation of life" numbers used by the insurance industry as far back as the Victorian era showed otherwise, as a *New York Times* article published on January 8, 1893 confirmed. "The value of outdoor exercise, with abundance of fresh air and a clear conscience, is amply set forth in a comprehensive table showing the number per 100 of various occupations that attain the age of 70 or more," the *Times* reported. Farmers at the turn of the century, before widespread mechanization, were second only to clergy in longevity, with forty out of one hundred expected to live to the age of seventy. By the 1970s, by contrast, growing commodity crops had become one of the world's most lethal professions, and had sky-high insurance premiums to show for it.

As farm families began to have fewer children, and fewer of the children they did have opted to work in a new, more dangerous setting, the reality of chemical-intensive, labor-saving farming inevitably infused 4-H. By 1978 synopses of educational films offered by the Illinois 4-H increasingly read like the decade's popular rural-scare films, *Deliverance* and *The Texas Chainsaw Massacre*. Simply by writing Ordie Hogsett, safety specialist, in

Mumford Hall, county and regional 4-H staff could be loaned *For the Rest of Your Life*, ("This award-winning film dramatically conveys the dangers to sight posed by anhydrous ammonia"); *Protect Your Livestock from Radioactive Fallout*; *Hands Off* ("This film shows what can happen if you don't use a corn picker correctly"); and other less-than-feel-good flicks with self-explanatory titles such as *Poisons in Your Home*, *Winter Survival*, and *Rising Waters*.

4-H eventually changed with the times, moving in the direction Martin had urged circa the 1960s by allowing for greater gender intermingling. Busy parents would find it more convenient to take both farm sons and daughters to the same meeting, Martin pointed out, adding, "a mixed group develops a more rational and normal relationship of the sexes." As 4-H evolved to hold more enlightened views on gender in the 1980s and 1990s, generation X parents like Angela Crock, who had once been frustrated by the organization's traditionalism, began to return. "Now things have changed," Crock tells me, "and my son and daughter are in the same 4-H club together where they can choose any project they want. My daughter is like me: she has no interest in sewing or indoor projects and would rather be outside on her horse. I love that they [my son and daughter] are both in 4-H; it teaches responsibility and organization. It's also excellent with getting them comfortable talking in front of a group and doing community projects. I think it's very important, and I am so glad our community has such a great club."

IN THE NEW millennium FFA and 4-H once more serve as revealing lenses through which to view the changing identity of the daughters and sons of rural America. While on the outside 4-H looks largely the same, offering dozens of animal-related projects for completion and exhibition, and a handful of programs in art and music, a closer look at its membership requirements point to a substantively altered world for today's rural daughters.

The first and most striking difference is the decline in club memberships since the mid-1960s, when *The 4-H Club Leader's Handbook* and other sources pegged the total membership in the US alone at 2 million. Membership requirements, too, appear to have loosened. While 4-H today claims to serve more than 6.5 million people, the actual numbers of memberships, according to the 2009 numbers from the National 4-H Headquarters Fact Sheet, struggle to top 1.5 million—a decline of over 20 percent from mid-1960s tallies. Just over 10 percent of 4-H participants, the Fact Sheet attests, self-identified as residing on a farm, while more than twice that figure, 23.6 percent, listed their residence as a city with a population of over 50,000. Moreover, to make 4-H more attractive to working parents and their children, membership expectations appear to have weakened. The Midwest 4-H member handbooks from the 1950s, for instance, required a farmer's daughter to keep an active record book of time spent on club activities and projects, attendance at two thirds of all meetings, and the delivering of at least one talk or demonstration to the club, beyond the required exhibiting of a project. By comparison, today's participation policy for Iowa reads somewhat more laissez faire. The section entitled "Who is a 4-Her?" defines a 4-Her as "an all-inclusive term which refers to kindergarten through 12th grade youth who participate in a 4-H program/activity designed to include a minimum of 6 hours of intentional, youth development experiences with a caring adult." Listed membership requirements do not require a monthly meeting or a set minimum record of attendance, but simply a "planned youth development learning experience throughout all or several months out of the year."

To check my perceptions against the facts, I got in touch with the youth program specialists employed through the extension service of my home state's land-grant university and my alma mater, Iowa State University, beginning with Youth Development Specialist Jackie Luckstead, whose territory covers Jones, Linn, and Benton counties in eastern Iowa. Luckstead opens by telling me she feels fortunate that she isn't covering more square miles, as are many of her colleagues in more sparsely populated corners of her once farm-dominated state. Luckstead volunteers her time as a leader of the Jackson Wise Owls 4-H club in Jones County, one of a dwindling number of groups who keep a monthly meeting. Luckstead admits that a series of state extension reorganizations and cutbacks have left her feeling spread thin, able to meet with 4-H members on average just once a month. She perseveres, however, because she believes in the communication and leadership skills the organization offers, and she abides by the historic 4-H imperative "to make the best better." Midwest girls, Luckstead's experience tells her, "need to be reminded that they can be the best."

"I have learned that 4-H has evolved into being not just a 'farm kid' kind of activity," says Luckstead, whose experience includes twenty years of working with 4-H kids and eight years as a 4-H girl herself in the 1970s, when she proudly showed hogs from her father's buying station. "There are as many or more youth involved in 4-H from the towns and cities as there are from the farm. Many farm youth are actually living on acreages." Farm girls living on self-supporting farms are few, Luckstead points out, estimating their numbers at perhaps 5 percent. The biggest change she's witnessed in her tenure is what she describes as "the city coming to the farm" to "take away many acres of farm production." This new, more urban population, she tells me, often can't wait to build their dream home in the country, though they soon find themselves "complaining about the smells, the dust, and the dark." Luckstead's most difficult challenge is trying to expand programming without draining what she calls the "biggest commodity" in youth leadership training these days: time. She argues that 4-H projects help in theory, but concedes, "There is not a good measure of how much the youth learn about the project, except to be able to explain the project to a judge during an interview."

Consistent with the 4-H strategic plan, Luckstead and her field specialist colleagues are endeavoring to incorporate more science and math, but that programmatic push seems to leave other curriculum areas spread thin. "I think that in this changing world, away from the working farm, we are trying to evolve toward teaching science and are offering more project-related club opportunities," Luckstead comments. "I do think that youth today are missing the farm-type experiences, and they are missing the family experiences of cooking and eating together."

Luckstead and a core of club volunteers plan to offer cooking classes in the small towns in her territory as a way of "filling in the missing pieces," though she admits, "I feel that there is a generation that missed learning how to parent, how to cook, how to garden, et cetera. These are the parents of the 4-H youth that we are seeing today." 4-H, she confirms, persists in teaching skills once thought "domestic" but mostly via project "hot sheets" and other resource materials available on the Internet. Learning about sewing, baking, and other domestic skills would depend, she says, "on the youth having someone that could teach them," and generation X volunteers with the time and ability to teach handicrafts viewed as old-fashioned are proving harder and harder to come by. "We have held workshops in the past and had very low attendance," she confesses. "Again, time is the factor stated as being the primary reason why the 4-H member doesn't attend workshops. The families are busy; the youth are busy. When do they find time to learn to live?"

The good news is that, despite pinched resources and shifting culture mores, data suggest 4-H girls and boys remain better endowed with life skills by comparison with their peers. A 2009 study by Tufts University confirmed that 4-H youth participants were more civically active than kids who participated in other out-of-school activities, while ninth grade 4-H youths achieved higher grades and were more likely to see themselves going to college. Depression and delinquency, the study likewise concluded, were more difficult to find in 4-H participants than in youth subscribing to other out-of-school programs. "Teachers say that they can pick out the 4-H kids in class," Luckstead tells me, "[because] they are not afraid to speak, probably from giving presentations and being interviewed through 4-H. They also have better writing skills from writing goals and doing record books."

"Statewide, 4-H is making very intentional efforts to reach the underserved," Daleta Christensen tells me when I ask her whether 4-H still teaches the very aspects of rural life—sewing, baking, cooking—Luckstead had suggested may be lacking. "As less and less youth live on farms and rural areas, we have strategically targeted urban environments in order to reach the majority of youth. As you know, all youth can benefit from the 4-H experience, which goes beyond project mastery and includes leadership, communication, citizenship, and other life skills," Christensen notes, adding, "We are making concerted efforts to reach an underserved population, urban youth, [and] minority youth."

I take Christensen up on her offer to take a closer look at the more than thirty projects my midwestern state's 4-H offers in the category of food and nutrition. The hot sheet, I discover, abounds with ideas for how food and nutrition might be learned. But as Luckstead warned, there's not much opportunity for a Midwest farmer's daughter to demonstrate true competence, let alone mastery. Today's 4-H farm girl is invited, for example, to "tell your family about the importance of eating a variety of food from all food groups," "design a poster about kitchen safety," or "create a club fundraiser around food." While such skills are important, they do not guarantee, or even necessarily imply, that she is able to prepare the meal about which she speechifies. Indeed, today's hot sheets appear infinitely less complex and nuanced than the projects suggested in Martin's nearly 50-year-old handbook, when a farm daughter might be faced with the thorny responsibility of planting, building, and maintaining a multiflora rose fence—a season's-long task for which success or failure would prove painfully clear.

While during the intervening generations 4-H has answered Martin's call to be "sufficiently elastic and progressive to meet the changing needs of developing youth," its very adaptability may have unwittingly sacrificed the skills that once distinguished Midwest farmers' daughters and sons, and the kind of year-long, rain-or-shine commitment my grandmother and the Pioneer Peps made so long ago. While *The 4-H Club Leader's Handbook* once spoke of making its female and male members "efficient, public-spirited and useful citizens," the front page of the August 2010 mailer produced by Iowa State University Extension mentions nothing of such high ideals, asking potential 4-H participants instead if they want to "make new friends that like the same things you like," "have fun on field trips with your friends," and "do cool stuff and hang out."

IN AUGUST OF 1935 Midwest farmer's daughter and member of the 4-H Nimble Fingers Club from Denison, Iowa, Donna Reed, arrived at the Iowa State Fair in a uniform she made herself—a middy with a pleated skirt. In a letter to her longtime pen pal Violet dated September 6, 1935, and addressed "Cher Amie," she conveyed the exhilarating events of the State Fair. "The last night I was there we 4-H girls, around 1200, marched down Grand Avenue to the amphitheater, singing the song 'Iowa.'" A handful of lines later, Reed picked up the breathless narrative. "Then I saw the agricultural building in which were flowers, fruits, and exhibits from every county in Iowa (99). By the way Crawford [County] received first. (Quite an honor to the County agent). Oh yes, my butterhorn rolls received third place in the state. My brother received $33 in prize money on his livestock."

More than seventy-five years later I travel to the same ag building that awed a young Donna Reed to witness the very displays of flowers and fruit the future Academy Award-winner raved about in her epistle. On the surface not much has changed at the Iowa State Fair. The ag building is still there, though it's now officially named the "John Deere Agriculture Building sponsored by Alliant Energy." The world-famous butter cow still entertains madding crowds from behind forty-degree refrigerated glass, just as it has since 1911. And as a special treat the one hundredth anniversary replica of the original brings a farm daughter and farm son back into the picture, standing either side of 600 pounds of low moisture, pure cream Iowa butter molded into the shape of a bovine and its child attendants. For the first time in forty-six years, a schoolteacher from West Des Moines rather than farmer's daughter Norma "Duffy" Lyon has crafted the buttery icon. Lyon, having raised cattle of her own on the family's dairy farm, knew whereof she sculpted. "She majored in animal science at Iowa State University, and she knew the dairy cow from an artistic and scientific standpoint," her replacement, 34-year-old Sarah Pratt, said on taking up the butter knife after Lyon's untimely passing in 2010.

Still, the State Fair marigolds and irises on display in the ag building have never looked lovelier, and the FFA ribbons appear blue as ever, collectively making the argument that excellence is still excellence, agriculture still agriculture, regardless of more superficial changes. And yet for all the pomp and circumstance, the inviolability argument seems more and more unstable.

In advance of my visit, I'd contacted Dale Gruis, consultant for the Bureau of Community Colleges and Career and Technical Education for the Iowa Department of Education. As the state's hand-picked FFA advisor, Gruis is the man to ask about the seismic changes in an organization once known for its historical continuity, a calling card it put on the line in a 1988 name change from "Future Farmers of America" to the "National FFA Organization." A year later the member's magazine, *The National Future Farmer*, also reinvented itself, dropping its customary title for the more abstract, amber glow of *New Horizons*.

Both post-ag crisis name changes, cynics noted, seemed calculated to distance the federation from the very word, *farmers*, that had long been its bread and butter. The change in sobriquet, defenders fired back, had everything to do with acknowledging and addressing FFA's—and indeed farming's—evolving demographics; even to maintain its decades-long clout as an industry player, FFA plainly needed more young women, more minorities, and more urban dwellers. In a brave new world of segment marketing, the anachronistic appellation "Future Farmers of America" no longer seemed to cut it, delegates voting in favor of the name change realized. "FFA for a couple of decades has pushed the concept of 'agriculture' rather than 'farming,'" Gruis explained to me from his Des Moines office. "FFA still represents Future Farmers of America, but we choose to refer to FFA as the National FFA Organization. Few people say International Business Machines; they substitute with IBM. FFA chooses to focus on all agricultural occupations by not saying *Future Farmers*."

A farmer's son from Buffalo Center, Iowa, whose family was forced from their dairy farm into off-the-farm employment during the ag crisis, Gruis takes a progressive approach to the dilemmas facing FFA. He believes, for example, that the vilification of organic farmers by hidebound food producers must stop. He understands that the hundreds of thousands of women and men who have left the farm for what he calls a "good job" ("good job" signifying, in his experience, health care and benefits) are doing so for their economic survival, not because they have anything against the self-reliance and proud individualism associated with self-employed farmers.

Gruis's comments reflect the notion that if you can't change the wind, you can at least adjust your sails, and the winds blowing through contemporary agriculture mean in-town kids now substantially outnumber farm kids, even in the Iowa FFA. "Students that grow up on a farm are an increasingly rare commodity," he told me, even in the high school agricultural education programs FFA sponsors. "However, they often are not much different than town kids. I suspect that compared to the old days farm kids do not have the same opportunities that you and I had. Some modern farm equipment is too large and too expensive to turn kids loose to operate. Fewer farmers repair their own equipment. Many livestock producers have employees to help operate the operation." Paradoxically, as farming has become more inaccessible to many young adults, and as the profile of FFA's farm youth has turned increasingly urban and suburban, the organization has become emphatically more female. "Nationally, roughly 60 percent of chapter officers are female," Gruis observed. "In Iowa, the number of female FFA members has grown significantly since the farm crisis of the 1980s. I have always suspected that many males were discouraged from agricultural careers, and ag ed instructors tweaked their curriculum to better attract female membership."

I'd asked Gruis if he could put me in touch with one of those promising young female officers, a farmer's daughter preferably, and he pointed me in the direction of Iowa FFA Secretary Jamie Leistikow, a young woman who grew up within the city limits of Readlyn, Iowa, but whose father helped the family cultivate 2,500 acres of corn and 300 acres of soybeans while paying the bills with the profits from his ag mechanic shop. As a farm daughter living most of her younger years off-farm, Leistikow reports that for chores she "hung around the shop, helped load junk and metal scrapings for extra cash, rode in the combines [and] tractors and semis when unloading corn, visit[ed] our pigs, getting dirty, and did little jobs to help my dad." Her mandatory FFA Supervised Agriculture Experience (SAE), taught her much, she says, and consisted of working in the ag room at her local high school keeping records and filing for her teacher, Mrs. Doese. Her other SAE consisted of "mowing lawn" at the family's three hog confinement facilities and "loading and unloading bait stations" around each building. "It didn't smell the greatest," she confessed, "but it was good money. Other random jobs included picking up rock in the fields during the summer and running errands for parts as well."

The well-spoken Leistikow wouldn't trade her FFA experience for the world, she'd mentioned to me, adding that her love of agriculture caused her to select a dorm floor at Iowa State University where half of her peers major in some aspect of the industry. However, as with so many of her FFA peers, Leistikow isn't banking on a farm future. "A lot of people are surprised that I'm a state officer because I'm not continuing my agriculture education, but I love what it has done for me, preparing me for the future in any decision I make. I am planning on going to California or New York for a future career in the fashion industry. My take on fashion might be different because of where I have grown up, and I am proud of it. For me, living in a busier town such as Ames, for example, excites me more since there are more activities to do than farming."

WITH THE MANY ironies of contemporary farm daughterhood still swirling in my head, I climb the stairs of a State Fair building that itself has been renamed to more transparently reflect the commercialism of contemporary food producers. On the building's mezzanine level I meet Nicole Patterson, FFA north central state vice president, and Shaniel Smith, FFA state reporter. They're both Midwest farmers' daughters, both 18, and in the Google-era, both in a tiny minority of working farm kids in their districts. "We only had a few coming from my school," Smith confesses shortly after we've begun our interview. "It just seems like there are less and less kids growing up on working farms, because farms are consolidating." I ask the girls what skills they had learned that perhaps their urban and suburban girlfriends lack. "Growing up on a farm has taught me to work hard," Patterson chimes in. "Everything that you do you are rewarded for—whether it be just income or people. When people talk to you about what you farm, it's just awesome. They're like, 'Oh! You *do* this?'"

"It teaches you good work ethic," Smith adds, "because if you're in the animal industry, those animals' lives are dependent upon you. So if it's 90 degrees out and 100 percent humidity or it's 20 below, you still have to go out there and feed them, even if it's in the middle of a blizzard. So it's work ethic and responsibility and you also get rewarded because if you bring a successful crop in then that's how you get paid." Does either plan to stay in-state and farm? Patterson answers first, the leadership training FFA provides well preparing her for questions like these. "That has been my ultimate goal since my freshman year in high school . . . to come back and eventually settle down on my family farm and raise a family around the area that I am originally from."

"I'm going to go to college for beef cattle science," Smith interjects, "and minor in agronomy. And I'm going to go back home after I get my college education and take over our farm. It just gets passed down over generations, so I'm next in line. So I'll go home and start a family and live on the farm, because if you live on a farm you truly have a better work ethic and morals and everything. I just think there's no other way to be raised."

Both Patterson and Smith agree that their experience in FFA has well prepared them for a career in agriculture, as well as helped answer the question of why ag matters. "The FFA program is excellent for . . . setting you up to go home and farm," Patterson says. "And along with that, growing up on the farm you see how things work and your parents are saying to you, 'This is kind of how we do things around here. So when you come back home . . . we'll hand it over to you.' So I feel like I will be perfectly ready when I get to that point in my life."

Smith has noticed the same modus operandi from her parents. "It seems like they start you out with the little jobs when you're younger, but the older you get they push more responsibility on you because they know you're older, and they're just preparing you to come back and take care of the farm when you get through college."

Is there a place on the farm awaiting Patterson when she's ready to come back home? "It all depends on the timing," she tells me. "If the timing is right, obviously I would hope to move into the home I grew up in on the farm. But my parents are still pretty young yet, so if they are ready to leave the farm and want to still stay at the house, I would maybe have to find a different place to live. . . . Actually, a couple of weeks ago I had the opportunity to go to the USDA in Washington, DC, and that sparked a huge interest in my career. And I wouldn't mind going to a bigger city and working for the USDA or something like the Farm Bureau." Smith agrees, "I would rather just come back straight home and go and farm, but if I get an opportunity to go to a bigger town for a few years, I might do it."

Happy as I am for Smith's and Patterson's teenage certainties about their vocational future—similar to those I held at their age—I can't help but wonder at whether they're prepared to resist the regional brain drain to metropolises such as Chicago, Minneapolis, Indianapolis, Kansas City, and Denver—the kinds of places that have historically drawn a near-majority of their peers for several successive generations. In his best-selling book *Who's Your City?* demographer Richard Florida argues that the massive rural-to-urban migration of the late nineteenth century appears likely to pale in comparison to the impending "spatial sorting" projected to create American megalopolises numbering perhaps a half billion people each. American rural places beyond the reach of these ever-expanding mega-regions—including most of Iowa, Missouri, Minnesota, Wisconsin, Nebraska, Kansas, and the Dakotas, are likely be left further behind as even the most sincerely repentant, college-educated young ruralites like Smith and Patterson join their peers in what Florida calls a "means migration" of capitol and education cityward. Increasingly, the best-selling author alleges, the ambitious and forward-looking will be compelled to leave the rooted class to join their mobile counterparts migrating to a handful of exclusively metropolitan areas where they are able to "realize their full economic potential." Demography, Florida writes, is destiny.

I thank the girls for their time and make my way further down the concourse, wondering whether Patterson really will return to run her parents' 3,000 acres of row crops and two hog production facilities, or Smith ultimately manage her family's 120-head cow/calf operation. Will Jamie Leistikow follow her dream of living and working in a fashion-forward metropolis like New York City? Certainly, my three interviewees are agriculturally endowed, the equity head start on offer in their families crucial in an industry that depends more than ever on inherited acres, equipment, and know-how. Still, the best demographic data overwhelmingly suggest that, despite their advantages, my FFA interviewees will not become future farmers.

At President Obama's Rural Economic forum, when state FFA president Lindsey Calvert proudly posed before a national press corps eager to photograph her as a seeming anachronism, the *White House Rural Council Report* predicted for farm children a less-than-rosy fate. Since 1960, the report indicated, the percentage of Americans living in rural communities had declined by nearly 50 percent. Outmigration among young people, it added, had "fundamentally shifted rural age demographics" and created "long-term challenges for job creation in rural areas." The report contained still more macabre news for Google-era Midwest farmers' daughters—if-it-bleeds-it-leads headlines likely to make them think twice about their dream of returning to their family's green acres. Rates of educational attainment had diverged significantly among rural and urban populations, the White House Rural Council intimated, so much so that an urban resident is now

10 to 15 percentage points more likely to have attended college than a rural resident circa the year 2000.

The cold hard facts say that Patterson and her kindred Midwest farmers' daughters are not likely to return to the farm after spending their 20s in a city like Chicago or Kansas City or Washington, DC. "Places that lose young people will never be able to recoup," Florida points out, "because moving slows down with age." Jamie Leistikow's first-choice field, fashion and apparel, for instance, points her in the direction of worlds far beyond the Corn Belt. Patterson's chosen profession, agriculture communications, is more likely find her behind a desk than at the helm of a combine. Even if she is able to superintend the farm while working full time in the agricultural press, her daughter, should she have one, would not be the daughter of a working farm, as Nicole herself had been. Increasingly, the question before not just FFA and 4-H, but before policy makers and rural advocates nationwide, is does that fact matter.

"A happy homesteader in front of her 'soddy.' The vastness of the country does not daunt her. She learns to love the quiet, broken only by the roar of a river at the bottom of a canyon or the howl of a coyote on the great sandy flats."

CHAPTER TEN

AG—VOCATING
WOMEN

"THERE ARE DAYS when divorce is just not good enough. . . . This is why we're in the jungle tonight."

After an hours-long autobahn on Interstate 80 to Des Moines, Iowa, I arrive at the Holiday Inn Heartland Ballroom in time to hear the first volley of keynote speaker Jolene Brown's address to the Farm Bureau Young Farmer Conference. Packed in around the dining tables are several hundred members of what appear to be, even in America's breadbasket, an endangered species: young commodity farmers.

After searching in vain for a seat, after begging the pardon of a young female farmer whose adjacent seat turned out to be occupied by a very large husband temporarily gone to the little boys' room, I crumple into an empty chair in front of a plate of fried chicken gnawed to the bone and a sagging piece of cherry pie à la mode, while onstage Brown segues into the backstory of her lead-off joke. She'd momentarily wanted a divorce from her farmer husband, Keith, after she'd accidentally backed their truck into a grain bin. Come to find out, hubby had jerry rigged a series of blocks to hold down the accelerator on the old, stall-prone pickup without telling his better half. Brown presumably deploys this tragicomic vignette both to assure the doubters in the audience of her farm bona fides (she's a full-time business partner to her husband, at least during harvest) and to reiterate the weekend's theme: "It's a jungle out there."

"There's going to be some slimy varmints along the way," Brown reminds the youthful Farmer Janes and Joes gathered in the cavernous ballroom. "And sometimes they live in Washington, DC." Laughter rumbles among the food-satiated as she continues, "Tonight is the night our celebration begins in the jungle."

Founded in 1914 and given an immediate boost by the passage of that year's Smith-Lever Act, the American Farm Bureau has not always been at odds with government bureaucrats; to the contrary, for much of its history, alleges agricultural historian David Danbom in his book *Born in the Country*, the Farm Bureau, in many rural localities, effectively "carried out policies" set by the government. The Young Farmer Conference that I'm attending provides plenty of evidence that the Feds and the Farm Bureau are still closely aligned where matters of farm policy are concerned.

Not so many weeks before tonight's dinner, US Secretary of Agriculture Tom Vilsack, former governor of Iowa, sounded the rallying cry for 100,000 new farmers. Here tonight, as the annual Young Farmer Conference kicks off, America's most powerful farm lobby has rolled out the red carpet for its next generation—picking up the tab for a day and night of speeches, meet-ups, dances, juried discussions, and seminars designed to empower young commodity farmers, a growing number of whom are women, with the strength to survive the slings and arrows of large-scale crop production. We registrants are plied with tickets for free drinks at the bar, offered complimentary child care at the adjacent Bennigan's Room, and titillated by a lobby full of pristine John Deere lawn tractors and Gators, any one of which might be ours.

In the ballroom the Bureau has worked its atmospheric magic, adding inflatable palms and stuffed jungle predators to reinforce the conference motif. In front of the attendees stands the six-foot-tall, bespectacled Brown, who commands our attention in her sequined gold blouse, reminding us of our favorite childhood aunt perhaps—the one who could be counted on to speak her mind, laugh and cry, and sometimes get in our face. "Farmer and Champion of Agriculture!" Brown's tagline in the helpfully supplied conference workbook reads. "Brown will have you laughing while you learn. She's fun and funny, long-legged but not long-winded, and so insightful audiences accuse her of sleeping under their beds."

"Tonight I want to take you to the world that's putting some pressure on us," Brown declares, prowling the stage. "We need to talk about how we are perceiving agriculture and our role." Tonight's speaker spoke recently at the annual meeting of the Associated Grocers in Seattle, and she's attending the Young Farmer Conference in part to tell us what we've already intuited—that things have changed in agriculture. "Now the four food groups are fast food, frozen food, dine-out, and carry-out. . . . We are not so much a have and have-not society, but a have and a have-right-now society." At her grocers' meeting Brown asked the managers of America's largest supermarkets to dish on where they made most of their money. They told her that superstores with drive-up windows that offer the convenience of McDonald's or Burger King but the healthful, choice-friendly smorgasbord of a deli are the real cash cows.

"After *my hard day at work*," Brown tells us now from the front of the room, lightly satirizing the prototypical urbanite, "I don't want to go traipsing up and down the grocery store aisle. But you know what I'm going to do? I'm going to go down the drive-thru and have roast beef with almonds. I could have mashed potatoes and garlic. I could have squash or peas. I could have a fresh green salad. I could have peach cobbler, and I could go home and have a home-cooked meal. This isn't my world, and I doubt if it's the world you live in either. But it is the world of the people who are buying our products right now."

"Many of you have grown a new appendage," Brown continues, building her case that it's not just food consumers whose culture has irrevocably changed, but the food producers'. "It's your iPhone or your Android. I bet it's not far away from you right now." Around the room, the mere mention of smartphones elicits involuntary pats of pockets to verify their whereabouts. "By the way, if your phone goes off during my talk, I'm coming around to collect $100 for your local 4-H, so you put that thing on vibrate, get it in your pocket, and get a cheap thrill."

Brown's slightly salacious humor is well-tailored to tonight's audience, young women and men who collectively grew up with enough exposure to artificial insemination and manure to appreciate good bathroom humor. Brown shares with us the story of a young woman in the adjoining stall whose phone interview rang while she and Brown perched atop adjacent porcelain thrones at the airport. "I didn't think it would be good for her if I flushed," she wisecracks. "We celebrated her new job while we washed hands."

"There's a world out there that doesn't understand what we do, and yet they have a big effect on whether or not we can do it," she intones, turning serious. "Why? They vote. They put together a proposition. Or they give money to the Humane Society of the United States, and then they tell you how to vote. Or they watch Oprah, Katie Couric, or Diane Sawyer, who now have 'fur children'—not pets, *fur children*—and those people don't know what we do."

We are less than a half hour into the opening speech of the conference and already the traditional enemies of the Farm Bureau—Oprah, the Humane Society of the United States—collectively "those people"—have been used to galvanize our agrarian pride and remind us that while midwestern Farmer Browns may know better than to believe such propaganda, the rest of the nation, surrounded not by farms, but by cities and suburbs, might not be so savvy.

Next come the stories of civilian-farmer encounters on the many airplanes Brown has traveled in a jet-setting life as a public speaker. Her seatmates, she says, are tickled once they realize they're seated next to "a real Farmer Brown," and a female Farmer Brown at that. "Boy, do we have some fun. Especially once they know that I can inseminate a cow. Believe me, these long arms have been put in places you can't imagine." Jokes aside, Brown says, her accidental encounters with American food consumers often take a dark turn. She maintains that the word "commercial farmer" consistently elicits accusations of "government bailouts," "murdering animals," "playing God," and "Frankenfoods." Her favorite dig, she avows, came from one confused yet indignant bystander who told her, "How dare you farmers make me eat those DNAs!" Other of Brown's ironic favorites include: "Do cows that give us skim milk drink more water?" And her all-time favorite: "When the farmers take the feathers off the chicken, do the chickens get cold?"

"I didn't want to say, yep, because they're dead and in the freezer," Brown cracks, concluding, "The people have changed. . . . I'm a lot like you, the more I deal with people the more I like hogs." In fact, our hostess tells us, the commercial farmer's image has been so thoroughly degraded during the years Brown has been crisscrossing the Midwest that she no longer admits to strangers that she is a farmer. She's in the "consumer service and products industry," she tells them. When inevitably they ask what she produces, she answers, "The food for your family, the clothing on your back, the fuel for your car. I am an American farmer." If she's feeling extra feisty, she asks, "By the way, do you breathe oxygen? I make it. I make it by the acres-full."

Brown's point is a serious one, and she chases the self-effacing humor by invoking the tone of a formidable farm matriarch. "Listen up, the value of what we do is in the eye of the purchaser, not the producer, in the eyes of the consumer, not the creator. Unless you and I can say what we do in terms of what they value, then we're going to be out of business." In sum, she says, young farmers would do well to think of themselves as "a walking, talking commercial," so they are able to counter consumer ignorance and answer consumer queries within the thirty-second time frame allowed before listeners shut their ears and close their minds. This very disconnect, she alleges, can be seen anywhere traditional farmers come together with ideologically armed consumers, and that includes not just the obvious crossroads like international airports, but also the very acres surrounding Big Ten university towns like Lincoln, West Lafayette, and Madison.

As an example of the agricultural ignorance of her increasingly urbanized neighbors, Brown cites a neighbor of hers who works as an auto industry engineer in Detroit fully 500 miles away. She only has to report to work every couple of months, allowing her, in Brown's words, "this little farmhouse that she bought and the 10 acres" and the romantic claim of a life lived in rural America. "She doesn't believe that a tractor should run after 7 o'clock at night. No farm equipment should take up more than half of the road. . . . Heaven forbid our neighbor in the pork industry should let his turd-hearse stop her when she goes down the road." Nevertheless, Brown's on friendly terms with this self-styled neo-ruralite, having invited her over to help with the harvest as an act of educational outreach, woman to woman. Still, every time the two share a flight back home from Detroit, she finds herself chagrined by the depth of her seatmate's ignorance of the land on which she purports to live.

"You folks sure do eat a lot of sweet corn," Brown says, doing her best impression of her high-flying, white-collar neighbor and seatmate—an imitation that comes off sounding like a cross between a suburban soccer mom and a namby-pamby Coastal foodie. "I said to her, 'Rachel, that field over there is disposable diapers. That field over there . . . you just drank it. It's the high-corn sugar syrup that they're using now in your soft drink. That field over there you just put in your briefcase. *USA Today* is printed on ink made from soybeans. And that fuel over there is going to take you home because you're riding with me, and we're burning a biodegradable fuel called ethanol."

How and if America's young female and male farmers accommodate an urbanizing world where a left turn into a rural driveway has turned into "one of the most dangerous things in American agriculture" is one reason why Brown, who daily lives such indignities, has been commissioned as the weekend's speaker. She wants the next-generation farmers in the room to honor their heritage and their culture, but not at the consequence of being eaten alive in the jungle. Her view of the issues facing contemporary farms is, in its way, wider and more grassroots than many university-based ag experts, who too often find an academic niche within the farm world and stick tightly to it, losing the forest for the trees, the cornfield for the ears. When she addressed the Associated Grocers in Seattle, she tells us, she learned that contemporary consumers aren't buying food as much as they're buying "time," "youth," "safety," and "an experience." They want their food to be fast, convenient, and easy. They want it to be safe and slimming, helping them to feel young again. They want to feel as if they know their farmer. And since the preponderance of grocery shoppers are women, it's women, the vast majority urban and suburban, who are set to determine the direction of agriculture. Brown intimates that these same grocers, when asked what is really demanding shelf space, responded "local, organic, and natural." In fact, Brown recently joined a marketing caravan across Canada dedicated to local, organic, and natural, and at each stop, she admits, her caravan spoke to a packed house. But when the media learned she was a commercial farmer, their questions turned from solicitous to accusatory, their tone suggesting Brown might be something nearer to pariah.

"I, too, have a very large garden," she told them then and tonight she tells us. "I have two chest-type freezers. And I work very hard every summer and fall to gather the bounty and share with my family and some of my neighbors just as the people here do. . . . I think we should honor both groups, and we should label our food well. But I have a question to ask you. What food do you think went to Haiti? It's not the local organic and natural producer. It's the commercial farmer who fed the world. But we don't need a war. Let us honor all segments of agriculture."

And that will have to be the evening's last word, as babies have begun to cry, and mothers have begun riffling through their purses looking for answers. After Brown bids us goodnight, the Farm Bureau staff reoccupies the podium, promising the following day will be full of break-out sessions on such trendy, age-appropriate subjects as young farmers and social media. Despite the progressive programming, in crucial ways this weekend's conference might as well have been staged in 1968 or 1975. We, the Midwest's "young farmers," are still overwhelmingly Caucasian; the majority of us have not arrived here by virtue of our own capital or entrepreneurial pluck, but on account of an "equity head start" inherited from our parents; most of us are still wearing the boots and Western plaids of generations past. And we're still more than willing to laugh at what Brown calls her "bathroom humor." In the forty-five-minute address just now ended, I've counted three references to sex—oversexed yellow pythons in mating season ("If you ever see one white belly side up, that's gonna be a female, and you'd better give her room because there's men on the way to pay her a visit."), a not-so-veiled allusion to overnight conference baby-making, and finally, a closing Ole and Lena Norwegian joke involving three-in-one oil used as a sexual lubricant (Punchline when Ole discovers Lena is pregnant: "It's a good thing we didn't use the WD-40").

Our evening's thoroughgoing animalism makes sense. The occasion asks that we think of ourselves as young agriculturalists trapped in a jungle. The evening's keynote positions us thematically as prey, baits us with the prospect of food and sex, and, at times, chastises us for instincts run amok. Still, it's the thinking animal in young farmers, the one possessed of native cunning counterbalanced by a healthy sense of superego and hyper-rationalism, that the Farm Bureau hopes to conjure this weekend, and to honor.

WHEN THE FOLLOWING morning I arrive to the farm wife-led session "Speaking Up to Address Myths About Ag," I am pleased to see women filling nearly half the seats.

"One day not long after her first child was born," a Farm Bureau impresario introduces the morning's farm wife-led break-out session, "Liz Nieman was watching an episode of Oprah that lit a fire within her. Fallacies and half-truths about the livestock industry were abundant. After taking five pages of notes on the episode, Liz decided she must do something about all the inaccuracies that seemed to flow so easily from the tongues of influential people. Today she will share with you the triumphs and pitfalls she has experienced over the past five years of ag-vocating for the business and way of life she holds dear."

As it turns out Nieman, like so many of her generation, is not the conventional Midwest farmer's daughter of yore, if ever such a phantasm existed. Instead, she is the product of the Reagan-era Heartland, a child of the 1980s, who, while born on a farm, soon found her parents' union riven by divorce and her father forced by the ag crisis to quit farming and retreat into a role serving the bigger farmers who had weathered the storm. While her father struggled to pay the bills working as a hog-hauler, hay-seller, and trucker, Nieman lived with her mother, who remained on an acreage where Liz could raise horses and participate in 4-H and FFA.

"Farm Bureau to me is like 4-H for adults. . . . It's the next step," she tells us now by way of entrée. Nieman claims she learned more about hands-on agriculture than many of her female peers in Chickasaw County, who, while they may have been born a daughter of the farm, gravitated not to-ward 4-H and FFA, but toward extracurriculars like volleyball, basketball, and spirit club. "Through high school I worked for farmers doing different things, picking rock, helping my dad deliver hay, doing bookwork. I always knew I was going to do something in agriculture."

While earning her associate degree in ag sales and service at Northeast Iowa Community College, Nieman worked office jobs for two different farm co-ops, where she performed a variety of tasks ranging from manag-ing accounts to delivering fertilizer, while concurrently offering her services to a large animal veterinarian. After graduating, Nieman found a position first with the Winn Co-op in Decorah, Iowa, before an engagement to her now husband Justin transplanted her to a retail job at a farm supply store in Delaware County. "I have a broad range of experience with farmers," Nieman tells us, summing up her all-important agricultural résumé to an audience known for tuning out "expert" speakers who haven't earned their pulpit with some serious sweat equity.

For all Nieman's experience, however, she is allowed to speak before us today not principally by virtue of her two-year ag degree or the breadth of her agricultural experience, but by marriage, a credential-by-association in-cluded for good measure in the impresario's intro to Liz as, "a full-time farm wife" who "presides over the Delaware County Farm Board," in addition to mothering three and helping her husband on the farm. As much as the Farm Bureau has evolved in the years I have been a member, it's doubtful Nieman would have been granted the podium if she were a single, child-less woman with a master's degree running her own organic farm. The in-stitution of marriage, and the principles of partnership, and fertility that bulwark it, remain at the heart of the Bureau's ethos.

"I became a full-time farmer in 2005," Nieman recalls, adding, "It wasn't too long after I married I decided to become a full-time mother. Ten months, in fact." Nieman smiles wryly, pausing a moment to let us do the math. "Looking to the future, I hope that one or all of them end up in farming or agriculture in one way or another. I'd be tickled pink if that happened."

Already, we like Liz Neiman. While her Midwest farm-daughter pedigree may be marred in some eyes by divorce and disunion, so, in large part, is ours. She's humble, easy-going, laid-back—like us, far from a confrontationist but able to stand up for herself and her family as need be. And that's exactly the side of Nieman the Farm Bureau has asked her here to talk up. "How many of you know this man?" Neiman queries, firing up a PowerPoint and splashing across the screen a mug shot that might as well be the Farm Bureau's "Most Wanted": the Humane Society of the United States (HSUS) CEO and President Wayne Pacelle—with his GQ looks and slicked back hair, a sitting duck target for conventional farmers if ever there was one. Nieman recalls how she was watching Oprah one day back in 2005 ("Ya, I know, I was watching Oprah . . .") and how she saw "this man" grandstanding against the so-called "abuse" of farm animals as he argued hard for an initiative, Proposition 2, designed to end it.

"They [USPS] are a huge fund-raising entity that plays off the name 'Humane Society.' There's not too much affiliation with your local humane society," Nieman explains. "People will call them up and send in their donation thinking they're going to help rescue a puppy or a kitty. Wayne Pacelle will take that money and spend a small fraction of it on rescuing a puppy or a kitty. The rest of it goes into lobbying efforts. . . . Wayne Pacelle is *a vegan.*" Our speaker stops to let the full cultural import of the damn sink in with our meat and potatoes crowd. "He had this cardboard pig up on stage with Oprah and this cardboard chicken and this cardboard 'cage' that a sow had to live in. He said they can't even turn around, these poor sows. They're abused. It's awful. It's terrible. Everybody in the crowd is like 'Oh, that's so terrible.'" While Pacelle held court for some twenty to thirty minutes, Nieman recalls, the token hog farmer Oprah invited was only given "two minutes of screen time, if he was lucky. . . . He was clearly worked up . . . who wouldn't be?"

Farmers all over America still blame Oprah for perpetuating the alarmist Mad Cow zealotry that nearly cost many their livelihoods, and it was enough to send people like Liz and Justin Nieman over the edge. "I looked at my son playing on the floor, and I'm like, 'This *man* is going to be the man dictating how my animals are raised, and he doesn't even have animals. How does that work?"

I presume Nieman's rhetoric is merely playing to today's audience, but a quick check of an online *Washington Post* article, "Vegan in the Henhouse," reveals she's right; Pacelle maintains he travels too much to responsibly care for a pet, though he hastens to add that he had pets growing up. The article, favorable in its portrayal of the youthful go-getter with the Kennedy good looks and just enough salt and pepper to suggest erudition, nevertheless quotes some of Pacelle's most ardent detractors, including Beth Ruth, of the US Sportsmen's Alliance, who calls him "Enemy Number One," and Patti Strand, of the National Animal Alliance, who evokes the "wolf in sheep's clothing."

On the farm as in the think tanks and lobbyist's offices of Washington, DC, the gender identities of America's strongest animal advocates have turned topsy-turvy. While women have traditionally been viewed as front-liners in the defense of animal rights, the HSUS, by far the largest and richest animal advocacy organization in the world, has had two consecutive male executives, while traditionally male sportsmen alliances and clubs have increasingly turned to female executives to offset a hyper-masculine public image. Not surprisingly, the same gender inversion has reached farm advocacy organizations, where 2011 finds the Farm Bureau's senior executive directors and executive assistant posts occupied by women, roughly 75 percent of its general counsel staffed by women, and well over half of its national public relations division populated by members of the fairer sex. Nieman, hand-picked to share her story with us in lieu of her equally young, equally fresh-faced, equally intelligent hog farmer husband, offers the most immediate case in point.

Nieman next cues one of the HSUS's most provocative commercials from the Prop 2 campaign. "Is this better now, or now," the commercial's voice-over reads, conjuring the classic optometrist's question as images of caged and abused factory farm animals alternate with the images of happy, free-range animals and their loving human guardians. "Number one or number two?" Nieman lets the campaign spot run through to its end. "Proposition 2 is a moderate measure that will stop cruel and inhumane treatment of animals crammed into cages so small they can't even turn around or stretch their limbs."

"How do you feel after watching that?" Nieman asks. "Anybody?"

"You shouldn't play with chickens . . . you'll get the bird flu," responds a droll young farmer from the back of the room, his arms folded defiantly across his chest.

Nieman, too, found herself sufficiently exorcised about the way others were representing her industry that she went on Oprah.com to defend herself. "I can't believe that you believe this stuff," she recalls as the gist of the comments she'd posted that fateful day. "I raise my animals in confinement, and we work our butts off every day to take care of them. A comfortable animal is a profitable animal."

"I thought I had a very good message there," Nieman recollects, hastening to add that she was sorely mistaken. "They said, 'You see, all you care about is your profit. You're all about your dollars. You don't care about those animals.'" Nieman reports that she spent days fending off the overwhelming numbers of agitated ag detractors picking fights on the site. The highly ideological impassionatos on Oprah.com consistently accused this Midwest farmer's daughter of playing God, of killing animals by her own hand. In return Neiman asked them what they would do if they had a suffering animal that was going to die. "They'd never answer me," she tells us, exasperated even in the retelling. "They just kept painting me as a barbaric person. . . . Finally I just had to back out of the conversation. There was no winning."

The moral of the story, she tells us, is to "think before you speak. . . . Think how people are going to perceive your words and choose them carefully, because no matter what you say, it will be twisted. The good thing is the Internet has a short memory. No one knew who I was, and no one's going to remember that conversation I had. . . . It was a good way to learn that lesson without damaging my reputation, I guess you could say."

"And the most important lesson is don't expect to convince the crazies. If someone is not going to be receptive to your message and return your dialogue with the same amount of respect you're giving them, it's not worth engaging them. It's a waste of time. I learned that lesson that day. . . . It really soured me on speaking out about agriculture because I was like, 'This is pointless. Why should I even do this? They're just going to beat me up and make me feel bad about myself even though I know that what I'm doing is the right thing.'"

While Nieman leads our group in an exercise about young farmer stereotypes, my mind wanders to the rich and often troubling ironies faced by the modern-day Midwest farmer's daughter from a conventional working farm, an identity, to hear Nieman tell it, one has almost to hide for fear of "damaging her reputation" or her business bottom line. And yet this reactionism, infused in a conference whose very theme is "It's a Jungle Out There," seems to be part of what needs shedding in order for farm daughters to be fully themselves again. It's exactly this conclusion that, after a few years of "laying low" and licking her wounds, Nieman tells us she reached herself. If Pacelle was out there advocating and educating and making a public figure of himself, Neiman figured she should do the same.

"I grilled my husband, Justin, because he's been a farmer his whole life. Why do we do what we do? Why do we give our cattle hormones, and how do we use our antibiotics and that sort of thing. I knew why, and I knew what we did was okay and safe. But I needed the reassurance and the back-up. You have to know why you do what you do. Otherwise, how can you defend it?"

Exhausting her husband's knowledge, Neiman networked further, talking to local vets and feed men. Scars newly healed from her first online scrape with animal rights activists, she tried Facebook, where she made a point of posting family-friendly images from her farm as status updates. She figured distant relatives in her family who questioned what she was doing as a Midwest farm wife and mother of farmer's daughters should be able to see for themselves the tangible evidence of the joy the farm life cultivated in her and her kin. "I am the face of farming, not this big corporate person in a suit punching numbers and telling his workers what to do," Nieman reminds us.

After going back online to post on forums "with a little more friendly atmosphere than Oprah.com," Nieman attended the 2009 Farm Bureau Institute, a year-long program bankrolled by the Bureau and consisting of a handful of two-days sessions. "I wouldn't be up here today if I hadn't attended," she avows. In one particularly influential session, Michele Payn-Knoper, the founder of the Lebanon, Indiana-based Cause Matters Corp., hooked Nieman with her pitch on becoming an ag-vocate via social media. Payn-Knoper encouraged the increasingly female membership of the Farm Bureau Institute to craft what Nieman calls a "pro-ag message."

"I thought it looked right up my alley to blog, to direct a conversation, and have a specific topic and write about that. I enjoy writing a lot," Nieman explains. Beginning with her first post on March 31, 2010, her blog

"Life as an Iowa Farm Wife" has sought to explain agriculture in its most basic terms, as she might to a stranger on the street in some American metropolis. Initial, infrequent blogs on planting season and other fixtures of the life of a farm daughter and wife have evolved into on-message, thrice weekly postings—one on the farm, one on the family, and one on a down-home recipe. Nieman admits that her blog reaches mostly friends and family members, still she adds that nearly half of those who have approached her in recent months have been strangers. Old-fashioned methods of outreach can be effective, too, for farm wives and daughters, Nieman assures us, asking her audience to consider volunteering to speak about their way of life at career fairs, chamber of commerce meetings, and schools, where, she says, "teachers love to hear from farmers." She encourages us not to be afraid to engage our in-town coworkers in conversations about the farm: "They're probably curious about things, too, and they might be afraid to ask." We're asked to consider hosting a farm tour or an open house. "Of course people love free food, so feed 'em a sandwich," Nieman exhorts us, "and they'll learn something."

AFTER ANOTHER QUINTESSENTIAL Middle American feed, we, the several hundred young farmers of this most farm-centric of states, gather once more in the Heartland Ballroom to hear Jolene Brown's culminating presentation entitled "Lions and Tigers and Family, Oh My! Pitfalls and Snares that Break up a Family Business."

While in our first evening together beneath the disco ball, Brown had presented her family farm as an emblem of traditional commercial food production, a calling card she had proudly defended on a Canadian marketing tour featuring natural, local, and organic foods, on this, our second day together, she mentions that she and her husband grow high-end "identity-preserved crops" and also have their own organic pet food business. She tells us that her own two Midwest famer's daughters, Calista and Miranda, have "no interest in operating their farm," the youngest happily employed as a pastor and chaplain at Yale University in New Haven, Connecticut, and the eldest teaching at the Eastwind School of Holistic Healing.

It's impossible to know how this news registers with the young farmers in the room, if it registers at all, slipped in, as it is, amid more emphatic talk about the pitfalls of passing the farm to the next generation. I myself have come to this weekend's conference not as a farmer per se, but as a farm writer and scholar, a fourth generation Farm Bureau member, one whose hay value alone would likely qualify me as a farmer in the eyes of the US Census of Agriculture.

In fact, as I think back on the young farmers' bios I've heard in the discussion meets and break-out sessions, Liz Nieman's "Speaking Up to Address Myths About Ag," and the weekend's various get-to-know-you and team-building activities, I realize that roughly a third to a half of this select group of youthful "agriculturalists" would likely not be considered "traditional farmers" at all by longtime Bureau members. I have met women and men who, shortly into our get-to-know-you sessions, admitted they drew their salaries not from the farm, but from careers in education, consulting, and ag services. In fact, querying her in advance of this weekend's activities, I'd learned that the very person responsible for the conference's surfeit of young farm leader activities, the Farm Bureau's Mary Balvanz, was not herself raised on a farm. With nearly 70 percent of American farms grossing less than $25,000, off-the-farm income has been a fixture in US agriculture for generations, and now here we were, the sons and daughters of hobby farmers, part-time farmers, farm organizers, contract hog producers, ag educators, extension workers, agronomists, and farm journalists—miles away from the image the public holds of midwestern farmers.

In her own family, Brown reveals, being smart means being clear-eyed about the family and business you have, not the one you wish you had. Of her own Midwest farmer's daughters, she tells us, "Neither of them will be operating our farm. . . . But we want it to continue. We do not want it turned into a housing development, which could happen in our area. So what do my husband and I do? We go to meetings just like this. You're going to find us at our young farmer mentor meetings." Our emcee admits that she and her husband keep a list of people who attend not just the Farm Bureau's young farmer meetings, but also the county corn-growers and soybean association gatherings. "And because West Branch is a small community, it's not hard to figure out how they treat their parents. We can

find out have they been in 4-H and FFA, and do they have their children involved? And since certain values are important to my husband and myself,
we ask do they attend church and are their children in Sunday school? We
actually have a list of names for our daughters to consider to rent to when
something happens to us."

And when the inevitable "something" does happen to Jolene and her
husband, farm daughters Calista and Miranda, one living about an hour
and a half away, the other 1,100 miles, have their marching orders—direc-
tives about which Brown is clear. Her once-rural next-of-kin will work their
way down the list of names their parents have supplied, not to sell the land
or to demand lump sum payment of cash rent—a financial arrangement increasingly impossible for deeply indebted young farmers to bankroll—but
to rent on a crop-share contract. Brown insists she has no problem telling
her daughters, "The legacy of the land is more important to us than your
inheritance." Moreover, her own girls are not to sell anything for at least
two years. ("That's how long it will take them to figure out their father's
filing system," she quips.) "You will get your inheritance not as the cash-on-
demand of death, but in terms by which the next generation on the land
can continue to be on the land," Brown avows, as if her daughters are right
here in the room with us.

The admirable sentiments Brown professes are exceptional in their
way, midwestern elders more typically dividing up the land evenly between
the children—girls and boys—with, at best, a time-limited trust to prevent a
knee-jerk, money-hungry sale. The realities of contemporary Midwest demographics seem to dictate that most of the erstwhile farm children living
in midwestern, if not East or West Coast cities, will "cash in" as soon as the
family trust is dissolved. Brown's advice, however, echoes my own father's,
who always said of the tricky business of farm inheritance: "fair is not always
equal." Brown maintains that Heartland farm elders who really want the
legacy of the land to continue will look for a proven manager, one who has
consistently invested herself in the farm, though given the Heartland brain
drain, this ideal heir is increasingly hard to find within a few hours' drive.

"I hope everybody in the family business starts as family labor. I hope
you have to scoop poop. I hope you got the junky equipment. I hope you
have to work all the weekends. When you labor well, you get to manage

well," Brown declares. The number one job of an agricultural leader who wants his or her legacy to continue, the weekend's keynote speaker tell us, is to replace himself or herself—a seven- to ten-year undertaking that often begins around age 50, when many farm folks suddenly turn serious about transitioning leadership and ownership. Until then, she says, they're often hanging on, making sure they have enough for themselves to live. It's about that time, when they find themselves overwhelmed by new technology and hampered by growing health concerns, that they often look to their kids, who offer the promise of youthful cooperation and what Brown terms "cheap labor." The trouble is, the kids may not want to be brought in at all, may resent having their skills used as a barter when with the same set of talents they might command a higher salary in the city. And by the time a parent reaches the ripe old age of 60 or 70, their farm children have often spent forty and sometimes fifty years acclimating to lives lived in college towns and metropolitan areas rich in amenities.

Marriage, too, can upset the applecart of a presumed multigenerational family farm. Whereas historically it has often been a cultural assumption that a farm wife would play a supporting role to her husband, rubber stamping his ideas and visions and schemes if in fact her seal of approval was solicited at all, changing marital and gender norms promise a new identity for the Midwest farmer's wife, who is more likely now to come from the city or suburb or college town than from the far-flung rural villages of yesteryear. She doesn't marry to be unpaid labor, but to be a full partner whose college-informed ideas are not only listened to, but often implemented. Brown asserts that where marriage is concerned, it's important that everyone be on the same page if they want the farm business to endure. She tells the story of a recent banquet where a farm father, nearing retirement, announced that his rural bachelor son had finally tied the knot. Brown had followed up with the question she inevitably asks of the soon-to-be farm father-in-law: had the young man found himself a keeper? "Oh, we've got a good one," Brown reports the farm patriarch told her. "'She's got a lot of new ideas." At which point, we're told, the wife elbowed her husband and whispered. "That's not what you said at home. At home you said, 'All she wants to do is change everything!'" Brown also cautions the region's young farmers not to believe the idea that the farm can support anyone in the family who wants to work on its acres. She points to the figures cited earlier in the morning by Virginia Tech's David Kohl, namely that the cost of living for the average rural family of 3.8 is roughly $65 to $68 thousand a year. "There are other ways to make money as you're earning the right to work on the farm," she reminds us in closing.

NOT LONG AFTER I return from the conference to my own acres, I call up farm daughter Michele Payn-Knoper, the speaker at the 2010 Young Farmer Conference sponsored by the Farm Bureau, and the woman who inspired Liz Nieman to begin ag-vocating. In addition to famously asking farmers to kickbox on stage to simulate the agility needed to defend themselves against ag-detractors, Payn-Knoper is one of the most prodigious users of social media in agriculture, having been named one of Mashable's Top Three Twitter Users of the Year in 2009. MPK, as she's known by her fans, founded #AgChat and #FoodChat, a weekly moderated discussion on Twitter, in April 2009. There her most popular chat, farmers answering consumer questions, generated 3.1 million unique impressions in three hours on a February day, her website claims. I contact Payn-Knoper not by tweet, but by phone, hoping she will shed some light on the rise of Midwest farm daughters as spokespersons in a traditionally male-dominated industry. Payn-Knoper is herself a certified speaker and principal in Cause Matters Corp., a company she founded in 2001 to focus on agricultural advocacy, social media strategy, grassroots marketing, and corporate sponsorship development.

We begin our conversation with the usual agrarian small talk—the unusually warm weather, the merits of the soils near her home in Indiana relative to mine in Iowa—while segueing naturally into a discussion of the farms on which we each grew up. Here, however, Payn-Knoper stops me. She prefers not to talk about the loss of her family's farm to bankruptcy; she tells me it's painful for her to dwell on the past. When I ask her to tell me about the farm she lives on currently—how many acres, et cetera—she's cautious, saying, "We typically don't share that, either. Let's put it this way, it's larger than 10 [acres] and less than 100."

In the new world of social media-based ag-vocacy, I'm learning, image is everything, and success in each of Payn-Knoper's chosen fields—dairy genetics and agricultural consulting—depends on establishing a convincing connection to working farms and farming. As an ag speaker trading on the mantle of "farmer" or "farmer's daughter" can mean the difference between a $5,000 keynote and a busted bank account. In her time as the regional director of the National FFA Foundation, Payn-Knoper sold over $5 million in corporate sponsorships; she, arguably more than anyone in the industry, knows the monetary value of a carefully controlled image.

Payn-Knoper's ability to capitalize on the most compelling aspects of her farm daughter biography accounts in part for her meteoric professional rise in addition to an impressive résumé that includes working in the Ukraine and Egypt on United States Agency for International Development (US-AID) projects. Like the farmer's daughter stereotype of old, Payn-Knoper is conventionally attractive, able to melt the heart of the most stoical midwestern pork or cattle producer. In an aging industry, she is, in her early 40s, comparatively youthful, closer to the age of the average midwestern grain farmer's daughter than wife. Like the farm girl tomboy of old she's perfectly comfortable around men and comfortable in her own skin, able to smile winningly at an uncharitable needle and give the needle back, as needed. She's also unusually open to new ideas, another farm daughter-associated trait, encouraging farmers to close the Twitter gap with their urban counterparts, who are twice as likely to tweet as their rural brethren according to a survey by the Pew Research Center. The study shows that the fastest growing Facebook segment is women over 55—within four years of the average age of the female farm owner, according to the USDA.

As a girl who bought her first pedigreed heifer when she was nine years old in 1982, the founder of Cause Matters Corp. never really stopped to consider what glass ceilings—or barn roofs—might lay ahead of her. "As far as the role of a farm girl," she tells me, "I never was really raised to believe there was any difference between what boys could do and what girls could do on a farm, aside from my big brother who liked to tell me I couldn't do anything." Payn-Knoper will tell you she's been treated well as a farmer's daughter by male mentors, professors, and colleagues at Michigan State, where she earned undergraduate degrees in agricultural communications and animal science, and in her career as an ag marketer and fundraiser, but there have been times in her working life where the gender rebukes have stung. "I was young, blond, and female," she recalls, chuckling ruefully at the memory, "and I was told to my face that I would never be successful selling semen because I was a female. That really left a mark on me because I knew cow families and I knew cows. I knew the customers as well as any man there did. It was very limiting to me, and I actually stepped out of the artificial insemination industry because of it."

"Young, blond, and female . . . some people would say those are three advantages," Payn-Knoper continues. "They created some major credibility concerns, I would say, in the last five years for a variety of reasons. I don't really worry about those anymore." The reality is farmers' daughters have "no choice but to translate" their special knowledge and insight, Payn-Knoper says of the professional hurdle before her—how to get conventional commodity farmers to talk to increasingly finicky, city-based consumers. "Since I started my business, I've very much focused on connecting the farm gate to the consumer plate. I think the advent of social media and some of my experiences working with a variety of different groups around the food plate have actually lended themselves . . . to reaching out to people beyond agriculture as I do with reaching people in agriculture," she tells me. In fact, in serving as a liaison between producer and consumer Payn-Knoper is only applying the skills she learned by necessity as a Midwest farmer's daughter in Jonesville, a small town less than an hour drive from both Battle Creek and Ann Arbor, Michigan. "I guess one of the first impressions I had as a farmer's daughter was I knew from a pretty early age that my little girlfriends didn't all love cattle like I did. I noticed that they didn't know that their milk certainly didn't come from the milk cooler at school or the brown cardboard carton, as I did. Sometimes there would be comments if I smelled like manure. I had to be conscientious about that."

Payn-Knoper isn't exactly sure why the keynotes at the Iowa Young Farmer Conference and other industry gatherings seem increasingly to be delivered by women, but she thinks it may have something to do with the feminine gift for community- and consensus-building. "I do see that women are as strong a spokesperson as any group that are out there because women like to talk," she explains. "It's been interesting. . . . I can honestly say, when it comes to the whole organic versus conventional debate, that females are better than males. I just think it's the willingness to have the conversation, and the open mind not to bash one type of production choice."

Even the most cursory look at Payn-Knoper's Cause Matters website confirms her basic thesis, that American agriculture must open up, must advocate for the health and safety of its food and the goodness of its culture if it wants to remain economically and culturally viable. In Payn-Knoper's girlhood, Midwest farmers might have been afforded their stoicism by captive markets, but now, she maintains, they can no longer be the strong and silent types, unless perhaps more verbal farm spouses or farm children are manning the public relations front on their behalf. "We in agriculture can blame misinformed consumers who are likely 3-4 generations removed from the farm," Payn-Knoper's website states, "or we can take a look in the mirror and realize times have changed. Modern agriculture is not all about business shifts in the production, processing, and selling of food; the changes have to also include the agriculturist's desire to reach out to the people who are purchasing our products and making decisions that will impact our businesses long-term. They want to hear from us, and agriculture deserves to have our side of the story told!"

While Payn-Knoper works with feminine grace both sides of the agricultural fence—consulting small, sustainable, and organic as well as large row-crop operations—in her work as a certified public speaker she's something less than a complete agricultural relativist. The ardency of her pro-traditional agriculture stance may cost her a few potential clients in natural and local foods, but her boosterism stands to pay still greater dividends in bookings made by beleaguered industry advocates who appreciate a woman willing to don her kickboxing gloves and duke it out on behalf of maligned farmers.

"I choose not to put the label of farmer on everyone who raises food," Payn-Knoper confesses to me. "I think Shaun Haney said it best, 'If you buy your seeds at the hardware store, you're not a farmer. You're a gardener.' Personally, I don't believe that people who have rooftop gardens are farmers. I think they are gardeners, and if they're raising food for themselves and their communities, more power to them. However, I will also say that is not equivalent to somebody who has 500 acres or 1,000 acres because one is a livelihood and the other is a hobby."

Payn-Knoper is the mother of a farmer's daughter, or at least a country girl (she resists calling her minor milking operation a farm, saying, "It's not exactly the typical small hobby farm, but it certainly isn't a full-functioning operation."). She is aware that she is carrying on a generations-long tradition begun by her grandmother, a Midwest farmer's daughter and wife in Ohio, and her mother, who grew up on a vegetable farm. While she admits she wasn't especially close with either of these would-be female role models, she realizes that by birth alone she has been handed an important mantle. Still, she's careful not to ag-vocate that heritage too zealously when it comes to her most precious constituency: her daughter. She admits, "It is hard as a mother to be away from our little girl. . . . I'm not so sure I have it any better or any more difficult than a typical working mom. It's just different. Something that I have come to terms with over the last couple of years is that my daughter needs an example of what it looks like to follow your dreams and needs an example of what it's like to be a professional woman. The first place she is going to look is her mother. I try to remind myself of that when I'm on my fifth airplane in three days and not getting home until 1 a.m. so that I can get her up at 6:30 a.m. I very much consider it an honor to serve agriculture, and I consider that I have been blessed with a gift. Honestly, if I'm anything, I should probably be using that gift."

To illustrate the point Payn-Knoper shares a story about a special moment shared with her own Midwest farm daughter. "I took her to an event in downtown Indianapolis last summer. It just happened to be her and me, and there was a booth there somebody had set up to talk about food memories and experiences. . . . I turned on the video camera in the little booth . . . and she sat there and looked at the camera and said, 'Food comes from farms. Farmers are wonderful people. And we should give thanks to them because they work really hard, and they take good care of their animals to give you food.'"

TO WITNESS JOLENE Brown win over a room of no-bull young farmers is to see the power an articulate, charismatic farmer's daughter or wife uniquely bears within the culture that produced her—the very same persuasion that Michelle Payn-Knoper capitalizes on to convince 65-year-old dairy farmers to kickbox in front of a convention hall like they're Bruce Lee. At the Young Farmer Conference, Brown asks all of us, some four hundred young farmers who probably haven't drawn a picture free-hand since the third grade, to illustrate lions on typing paper with oversized crayons.

"I would like to analyze you now based on your lion and your future on the farm," Brown calls out to us while prowling the stage, pausing dramatically to ramp up the drama as only a talented raconteur can. Our right-facing lions, we discover, belong to the "visionaries" among us farm daughters and sons; left-facing lions signify grounded skeptics. "If your lion is facing you, listen up," our keynote speaker shouts over the growing hubbub in the ballroom, asking for a show of hands that includes my own. "You have issues that need to be addressed."

At the lectern she cracks a wide, wry grin, while we, the farmer-drawers of front-facing lions, turn redder than a roomful of cherry pies. "But you will do so because you are determined. You are bold. You like a challenge and you are going to achieve great things," she adds, redeeming us wholesale.

"This photograph of a Camp Fire Girl shows the opportunity country life affords for good sport."

CHAPTER ELEVEN

COMMUNITY–SUPPORTED AGRICULTURE

IN 1984 A prospective kindergartner teacher named Robin Van En moved from northern California with her young son to Egremont, Massachusetts, looking for a little land on which to build a new life. Van En intended to finish her teacher training, but instead she found herself investing in a 65-acre retired dairy farm in the Berkshire Hills. At first blush the location struck the new owner as symbolic. Johnny Appleseed was rumored to have planted his trees along nearby Jugend Road, and Shays' Rebellion, in which farmers famously revolted against government taxation, had taken place just over the hill. If these two portentous calling cards weren't enough, the famed Appalachian Trail and the first community land trust were located just the other side of Jugend Mountain.

That a young mother and teacher would inaugurate what is widely considered the first community-supported agriculture (CSA) in America by offering shares of their 1985 apple crop seems somehow appropriate. In fact, it had been a group of Japanese mothers in the mid-1960s who began the movement, many scholars contend, when they banded together to purchase fresh milk in an arrangement known as *tekei*. "Literally translated, *teikei* means 'partnership' or 'cooperation,'" Van En explains in the introduction to her cowritten book *Sharing the Harvest*, "but according to *teikei* members in Japan, the more philosophical translation is 'food with the farmer's face on it.'" Indeed CSAs have become synonymous with lending a face, and a profile, to community-based agriculture, and they were founded in opposition not only to the image of the corporate farm, but as an antidote to the image of the disconnected commodity farmer producing Frankenfoods for unknown consumers often living half a world away. In 1986 only one other CSA operation beyond Van En's was in existence according to USDA figures, while today's count of CSAs tallies more than 4,000. More numerous still are the farms the USDA counts as marketing products through CSAs—a number that topped 12,500 circa the 2007 Census of Agriculture.

In the early 1970s my own traditional, corn-and-beans farm family began to seek creative ways to transcend our conventional commodity crops via what a 1972 article in the *Iowa City Press-Citizen* called a "country market." The idea was simple: gather together artisans of both food and fiber for an old-fashioned, on-the-farm swap. In his 1946 book *The Furrow and Us*, my great-grandfather, a Quaker conservation farmer and early organics advocate, had written, "The well-rounded farmer is not a wage slave but a creative worker," and it was these words, in part, that inspired our community-based initiative. "We're doing this because we are interested in this type of community get-togethers," my grandmother Julia told the newspaper reporters when, inevitably, they came calling.

IN DECEMBER OF 2010 an exceptional article appeared in the Iowa Farm Bureau *Spokesman* headlined "Social connections key to reaching women farm owners." The *Spokesman*—with its legislative updates and crop reports and ads for genetically engineered corn and soybeans—is not what most would consider a women's magazine, but here was potential proof to the contrary, evidence of the new clout female farmers wielded among traditional commodity producers. The article opened with a shocker—nearly half of Iowa farmland was now owned by women. The key to reaching women in agriculture, a study by the Women, Land, and Legacy (WLL) project had found, was "social support"—in a nutshell, making farming more relational, more female, again.

"Social support is essential to the risk management strategies that women use to understand and validate their experiences, gather information, and gain confidence," Lois Wright Morton, Leopold Center interim director and Iowa State University sociology professor, said in a press release, adding, "Our work with women landowners needs to strengthen the connections women have to each other, the land, their families, and the providers of resources that are available to them." Beginning in 2004, WLL sent out twenty-three teams to host listening sessions for female farmland owners in thirty-seven Iowa counties. Among the team leaders was Laura Krouse, owner of Abbe Creek CSA in Mount Vernon, Iowa, and author of the 2009 report "Women Caring for the Land."

"I've done a lot of work with elderly ladies primarily with conservation education the last two winters . . . and those ladies are kind of the last ones who know what conservation is supposed to look like," Krouse tells me when I call to ask the former Cornell College biology professor turned community farmer what we might expect for the future of farmers' daughters. "When the older ladies die, their tenants and their children don't necessarily have in their heads what we're shooting for," Krouse explains. "It's going to be a big loss. . . . All this land is going to change hands, and it's going to people who have a real disconnect and don't seem to have the memory of how much good conservation requires."

These 80-year-old farmers' daughters know something their baby boomer children may not, claims Krouse. "A lot of my ladies were alive during the Dust Bowl, and can remember what it was like when it was so hot and so miserable. . . . They also remember smaller fields, smaller farms, closer neighbors, places that were safe for kids to play. They remember the farm as a functional part of a functional community. And they're willing to do some things to keep it as much like that as they can, rather than have it be just another 500 acres to add on to somebody's 10,000 and sell the house off to somebody who works in town."

"In 'Women Caring for the Land,' what we were trying to do was educate women," Krouse continues. "Because sometimes they, the tenant, is
now the grandson of their husband's best friend that died forty years ago,
and they don't know this kid, and the kid doesn't know them, and they're
afraid to talk to him, or he can't be bothered, or he's afraid to talk to them—
all reasons why they might get disconnected from their tenant. We thought
it was kind of helpful to say, 'This is your land. You own it. You get to make
the decisions.'"

By 2006 Krouse and the members of WLL planning teams had heard
from over eight hundred women attendees of their listening sessions. As a
follow-up she contacted district conservationists in the counties in which
she worked. Each one informed her in the weeks after the session that they
had been contacted by women landowners they had never heard from before. Beth Grabau, Farm Service Agency (FSA) public relations specialist
and one of the authors of the WLL report, confirms Krouse's assertions—
more women, she notes, really are stepping up to demand their share of
state and federal assistance. Grabau claims that FSA programs are equally
open to both sexes, but that fact doesn't always mean participation in those
programs turns out to be equitable. "I think one thing that is challenging
for women is that they are more likely to have an interest in a form of agriculture that is not traditional," Grabau says when I ask her about the rising tide of female agriculturalists, "so meeting those needs can be difficult,
as our programs are not always geared toward them. But that is changing.
Sometimes there is a steep learning curve for [FSA] employees as we have
to learn about that type of crop or enterprise and then try to fit programs
to meet their needs." In one sense the very idea that farming has become
so thoroughly programmatic, if not bureaucratic, Grabau suggests, may be
foreign to the farm daughters of a more self-reliant era. Consider Grabau's
own case. "We had corn and hay ground; at times we rented other hay
ground," she recalls of the family farm of her girlhood. "We picked corn
on the ear and would grind it all for feed. I remember shoveling corn into
the corncribs at times. I spent time baling hay as well. One did not say they
were bored, as there was always work cleaning out the barn. My mom had
a garden and canned most everything. She would can a minimum of 100
quarts of just tomato juice. She . . . made jams and jellies, we churned but-

ter, processed our own beef. . . . I am tired just thinking about it. We would probably be considered sustainable by today's terms."

"Women talking to women is critical in almost anything you want to accomplish," maintains Krouse, whose modus operandi for flushing out farm daughter landowners for participation in WLL was simple but effective. First she looked at tax rolls to find female landowners who appeared to be sole owners of large tracts. Next she took care to invite her target invitees to meetings slated for "neutral spaces" such as libraries or churches, while simultaneously undertaking to invite district conservationists and model renters to share best practices. Finally, she packed her ladies in a van and toured them around the countryside, pointing out conservation, or lack thereof, on the ground. "We would go by like ten million dysfunctional waterways, and we would stop at one and say, 'Ok, this is not how it's supposed to look. . . . If you look here, you can see the water's running down the edge rather instead of going into the waterway. And for a lot of women it makes perfect sense. Of course the water is supposed to run down the middle of the waterway. And they looked at that and said, 'Oh, mine look like that, too. I didn't realize that that's what that was.' All it really took was a conversation."

"I found one guy in every county who was well recognized as a good conservation tenant," Krouse explains. "We picked guys who are soft-spoken, and we had them come and talk to the group about how they interact with their landladies and landlords and then answer questions. I said, 'Pretend this guy is your tenant and you want to ask something. Practice.'" For some farmers' daughters in their dotage such a trial run was the first face-to-face business talk they'd ever initiated with a man as an equal negotiating party.

Krouse headed one of twenty-three teams that canvassed the Corn Belt, reaching an estimated fifteen hundred female landholders. Once the results were written up by lead author Corry Bregendahl at the Leopold Center for Sustainable Agriculture, a picture of a female-centered agriculture began to emerge. Not surprisingly, its needs and wants looked different from the male-dominated world of midwestern corn farming. Among the study's principal findings, women, more than men, "invested in relationships with others to help them make informed decisions about their farming operations" while "relying on kinship and friendship networks to help them make decisions about their land and agriculture." Women, the WLL project discovered, were much more likely to attend informational meetings in "peaceful, neutral spaces" where other attendees and presenters were women, and where there was an emphasis on storytelling and face-to-face exchange. They would also be more likely to bring their expertise to the table if child care, mentoring, and respect for their farm schedules proved forthcoming. WLL's conclusions, while self-evident in many regards, had potentially profound implications for Midwest farmers' daughters young and old. Read holistically, they suggest that women feel more comfortable owning land, farming land, and passing down farmland to the next generation if along the way they are mentored by other women.

Retroactively applied, the implications for the farm daughters of the 1970s, 1980s, and 1990s are intriguing. Perhaps country girls left the farm not because they were dissuaded from entering agriculture by sexist parents, not because they were naturally inclined toward urban settings, and not because they wanted an easier, white-collar life, but because agriculture hadn't been feminine enough to compel their interest. Had agriculture been more comfortable and comprehensible to concerned women, perhaps the generation of farm daughters that abandoned the homestead in the 1960s and 1970s would instead have voted with their feet to remain, keeping intact the cultural chain that, once broken, finds the children of one-time farming clans hiring management companies to superintend family farms from faraway cities.

AT ABBE CREEK CSA Laura Krouse incubates an agri-model consistent with the findings of the WLL project. She counts on women, and moms especially, whom she describes as "her best allies" in the struggle for safer, healthier food, to help her business grow. While she's careful not to discriminate, she asserts that young women make superior employees where community-supported agriculture is concerned. "I almost only hire young women. . . . And every young woman who has worked here has toyed with the idea of 'I'd like to work on farm, I'd like to live on a farm, I'd like to own a farm.'" Over the last three or four years, she's had three or four young women set out on their own agrarian path and attempt to make a go of it. None of them have been from farm backgrounds. "I think our future is with suburban young people who care a lot about the environment. They care a lot about food, they're very smart, and they look to farming for its intellectual and physical challenge. Together it's a package you can't get in very many other occupations. The really smart ones can see that this could be something to keep their attention for a long time." Kids who grow up on traditional grain farms don't usually come to work at Abbe Creek, my interviewee explains. Krouse hired a farm kid once, but the minute he was old enough to drive his dad's combine, he was history.

Krouse's own story parallels the young women's who apprentice in her greening fields. Like most of her disciples, Krouse didn't have family land waiting for her when she graduated from university. But unlike her youthful charges, she didn't have anyone in her life assuring her that her dream was possible. "When we were seniors, we had to take aptitude tests to pick a major and a career, and I was the last one that the counselor called in. He called me up to his office and said, 'I don't know what we're gonna do with you.' And I go, 'What do you mean?' And he said, 'Well, you came out plumber, and you can't be a plumber. What do you want to be? A teacher or a nurse?'"

Krouse had an inkling even then that food production—so closely tied to her aptitude for problem solving—would be her calling, so she chose a major in agronomy at her state land-grant university that would teach her everything she had yet to learn, and would require her to intern in areas about which she knew precious little. In college she took every hands-on internship and trainee position offered her, detasseling here and working at a fertilizer factory there. She threw herself into applied agriculture so furiously that she often didn't make it to class. She confesses, "I wanted to be a farmer, but then not having any land at home to go back to, I expected I would marry a farmer. That didn't happen."

Instead, cap and gown freshly pressed, Krouse shipped out to western Kansas to work at a co-op in Dodge City of "get-the-hell-out-of-Dodge" fame. From there she moved to Florida to do missionary work among the migrant farm laborers in Immokalee, a tomato and watermelon town, before pursuing a master's degree in agricultural extension work at the University of Florida. After years of fieldwork and consultations on farms across the Panhandle, she realized she'd amassed the knowledge to *do* farming rather than merely study it. After making $35 a month for seven years as a missionary and picking tomatoes for grocery money, however, she didn't have sufficient capital to buy even an acre of good midwestern dirt. She figured she'd take her newly minted master's degree and pursue agricultural development in Latin America. Life, though, had other plans in store for her.

Krouse recalls, "My mom found a little farm for sale near home, and said, 'Why don't you come look at it?' With the help of a high school banker friend and a parental co-signature, she bought her midwestern acres in 1988—in a drought year in the depths of the farm crisis. The first years were rough, but the farm scholar turned farm entrepreneur got by with equipment borrowed from neighbors, and an old combine bought for next-to-nothing at a sale. She recollects, "We hobbled along on broken and borrowed and old. But there were farm sales every weekend in '88." Though she was also working full time as a credentialed soil conservationist in a nearby county, as a single woman she was doing what at that time seemed impossible: practicing sustainable, shoestring farming in the era of big ag.

"I thought the farm could pay for itself and it did," she recalls. "I've been able to make the farm payment every year from the farm income." Most of her fortunate-son classmates from her university days weren't so lucky, she tells me. "In my ag business classes I was one of the very few people whose father didn't own a bank. I was in classes with very rich guys from very wealthy, big farm families who were investing in all the latest technology. As far as I know there aren't 10 percent of them still farming. . . . Those guys entered in the late '70s and early '80s, the absolute worst time. Interest rates were the highest. Land prices were the highest. So they had a lot more on the line than when I came along and cleaned up the wreckage."

Krouse chalks up her unlikely success to "good luck," but, viewed in another light, the success of Abbe Creek CSA offers a look at what may have been possible for other Midwest farmers' daughters with a burning desire to cultivate, but, like Krouse, lacking an equity head start. A supportive family, some timely university agriculture classes in business and agronomy, a willing hometown banker, thoughtfully nurtured relationships with neighbors, and a lot of determination can help a Midwest farmer's daughter, then as now, hoe the long row toward ownership. Still, it wasn't easy being a female farmworker breaking into the field in the late 1970s and early 1980s. "I had real bad things happen on jobs that shouldn't—that now you could get arrested for. And I was left out, lots of times, from the good jobs," Krouse says of those heady early days. "I think it kind of whizzed right past me though because I was much more interested in hands-on experience. When I left school I wanted to go West, and I wanted to learn the grain-trading business, but I didn't want to start in a grain-trader's office. I wanted to start on a dock or at an elevator because I wanted to start where the grain started. . . . I was one of the few girls in the trenches at the time."

"It [farming] was thought of as boring. It was thought of as inappropriate work for girls," Krouse remembers of the reasons why her fellow female classmates didn't follow her into the business. "I sure didn't know how to do anything when I went to my first job." She cites one encounter with a kindly farmer whose encouragement both confirmed her career choice and explained in hindsight why many daughters of working farms had long since left the fields for greener professional pastures. "I was at his place, and the chain on the emptying auger on the feed truck broke. I was standing there looking at it, and I'm like, 'Now what? I don't know how to fix it. I'm going have to go in the house and borrow the phone, and I'm going be late on my next load. I'm going to get yelled at.' And he came out and said, 'The reason you don't know how to fix this is because no one ever showed you. Boys are not born knowing how to fix roller chains. They know because somebody showed them. I'm going to show you, and you're going to be able fix them the rest of your life.' He did that, and now I know how to fix a roller chain. That was probably one of the most important things anybody's ever said to me in my life, actually."

When she considers potential hires each spring, Krouse goes fishing for exactly the skills WWL listening posts revealed to her as women's unique skill set. "When you're out here all by yourself and something goes wrong, you've got to know somebody who can help you," she explains. "Collaborative practice, that's huge, and I think it's definitely more likely to happen with women." Even in Mount Vernon, Iowa, a small college town of less than 10,000, the popularity of the CSA movement means Krouse is able to find seasonal female help from a backlist of eager candidates. She's hired young women who have been willing to drive from the university city twenty-five miles away to learn the skills she has to teach. "Ten years ago I had a hard time. Now I pretty much get to pick the very best girls. I want people who are real smart, and who are real confident, who aren't afraid to go try some stuff, take some leadership. I am at the place, the business is at the place, where working on an organic farm is cool."

After working with young apprentices on Abbe Creek for more than twenty years, Krouse is hopeful regarding the next generation of midwestern young women vested in agriculture. While she worries that the mostly urban and suburban females enrolling in sustainable ag apprenticeships, internships, and institutes may lack the native understanding yesterday's farm daughters were blessed with by birth, she's open to the possibility that with new blood may come new strengths. "We might get better, more thoughtful farmers out of the deal," she concedes, "rather than people who just do what they've always done. The thing we lose is community and how rural communities live with each other. A new person might not know the importance, the necessity, of showing up at the fireman's pancake breakfast, even though you don't like pancakes, because they need the money, and you need to see your neighbors. . . . New people in town who come from suburban areas don't realize that in rural areas somebody has to be the volunteer fireman, somebody has to be on the ambulance, somebody has to mow the cemeteries, somebody has to be on the committees. They've always lived in a community where that was all taken care of, and they don't realize that you have to step up in the rural area because the responsibility of citizenship is more demanding. Plus, carrying on the traditional things is necessary, too."

While, for example, clubs aren't necessarily Krouse's cup of tea, she sees what Robert Putnam perceived in his book *Bowling Alone*—that club culture, however old-fashioned it may seem to 20- and 30-somethings—helps grease the wheels of communal life in farm towns. Consequently, absence of those same values often implies slippage. "The young women don't come [to club] because they're at work, but the older women have been going their entire lives," Krouse says, citing a familiar generational gap in the Heartland. "Their whole support system is the women who are in club, who are the women who were their neighbors. When they got married and moved to the farm, they met these women who also were getting married and moving to the farm, and they, the older women, taught the younger women how to be a good neighbor, how to be a wife, how to be a mom, how to be a farmer. Whatever they had to learn, they learned it from these women . . . that's kind of falling apart."

While Krouse may be more liberal in her politics, she has learned to be appropriately conservative in her views of community. She acknowledges that she lives in a region and a state where corn is king, and anyone deviating much from that path, especially female food producers, is subject to intense scrutiny. She accepts these social constraints, and in her own way, honors them. "You can't go your whole life resenting that your neighbors know what you're doing and everything you say. You have to accept it and run with it. It might be harder for kids who didn't grow up in rural areas to allow themselves to be comfortable in that position. . . . In my case it's kind of a fine line. I want to be progressive to my progressive customers, and yet I want to be friends with all my neighbors. It's kind of a balancing act. I think the most successful women CSA operators are the ones that become a part of their communities. They serve on committees, they go to church, and they go to the potlucks and eat food that they would not cook themselves, and they don't gripe about it, and they don't bring tofu to church potlucks. They try to be part of the community as much as they possibly can. . . . I think that's very important. . . . For me it's been easy to take some pleasure in learning how to live with people, and I've never felt comprised doing that."

I WAIT FOR Bronwyn Weaver, honey-maker, beekeeper, and CSA CEO, in a kitchen where a group of a half dozen high-energy, granola-charged 20-somethings make their employee lunches. It's a brisk day in a belated growing season in La Fox, Illinois, a north wind howling and highs expected to do well to reach 60 degrees.

While I wait, I pick up a copy of the March 1977 *Organic Gardening* from the end table and commence to browsing. Inside the issue with the cover story "Fish Gardening Is Coming of Age," run 35-year-old ads that might as well have come from today's editions of *Mother Earth News* or *Country*

Living. "Millions dream of a tasty tomato in vein," one advert reads, talking up a lightweight, rust-proof tomato cage. In another advert Jeanette Zalucky enthuses about her tiller, "It's not just a matter of saving money on food. Our garden is basic to how we live and bring up our children."

I am lost in the amber glow of the back-to-the-land movement when Weaver walks in, greets me with a firm handshake, and invites me to have a seat in her fishbowl of an office, in front of which passes a steady stream of the CSA's youthful employees. Weaver—"Bron" as she's called for short—is a no-nonsense farmer's daughter turned shrewd head honcho at Prairie Heritage Farm. She is well-educated, vigorous, savvy, and emphatically nobody's fool. Since 2010, when Bron's Bee Company rolled out its chef-sponsored apiary program to much media ado, Weaver has become a major player in the Chicagoland natural foods scene, her business featured in both the *Chicago Tribune* and the *Chicago Sun-Times*, among others. Indeed, sponsoring one of her remote hives has led some of city's most prestigious chefs, including City Provisions' Cleetus Friedman, to almost child-like epiphany, reported the *Chicago Tribune* in November of 2010. "The first time I butchered an animal, I thought, 'Every chef needs to do this,'" Friedman enthused to the *Tribune's* Jennifer Day. "I thought the same thing when we did this. I can't describe what it's like to get up in the hive with thousands of bees flying around. It gets me teary-eyed thinking about it: These are my bees. They did this for us." Though the bees more likely did it for themselves, Friedman's quasi-religious experience is one Weaver cultivates, arguing to Friedman, Day, and anyone else who will listen: "There's just as much magic in honey as in heirloom tomatoes." Weaver has discovered that the revelations implicit in her small farm can induce grown women and men to schoolgirl or schoolboy wonder; to some she has become both mother and fertility goddess, able to understand if not command primal forces that leave the city's most knowledgeable chefs agape.

For Weaver, an Ohio farmer's daughter, the ecstatic experiences of the farm are old hat, though it hasn't always been that way. She describes her Ohio rural girlhood as "anything but a cookie-cutter experience." While members of her mother's side of the family worked as farmers and orchardists for generations, Weaver's parents initially followed the course of many farm-born kids, leaving the land for urban careers and suburban comforts.

But early in her midwestern girlhood Weaver's folks made what she calls a "right-angle change" and bought a 160 acres—a full quarter section—in the hills of Ohio. There the already thin soils had been farmed out to the point of erosion by the time 7-year-old Weaver and her parents arrived to sow the seeds of a sustainable farm and a new life together. Weaver remembers a neglected landscape tended haphazardly by weekend farmers holding down full-time jobs in nearby cities like Canton. Even that catch-as-catch-can Saturday cultivating had pretty much slowed to a trickle by the late 1960s, when the fallow Ohio foothills gradually washed away. That's when Weaver and her kin set up shop on the northern edge of an impoverished strip mining region, resolved to hold down the soil even as they held down the fort.

From that point forward Weaver would live the life of a farmer's daughter in the middle of what she called "an organic experiment" that included such novelties as accompanying her mom to the butcher shop to obtain blood and bones for the compost pile. In addition to raising organic crops, the young farm family raised cattle and chickens and hosted hives for honey bees. The whole clan bought in, Weaver's older sister sufficiently moved by her own farm daughter upbringing to pursue a degree in agricultural economics from The Ohio State University. "It wasn't that they were on a radical mission," Weaver tells me of her parents. "It was really them getting back to their roots. And that was this wonderful way that I was able to grow up."

Granted, Weaver was not a typical farmer's daughter. By third grade she was picking up the family's copy of the *Wall Street Journal* from their rural route mailbox, and reading it out loud to her mother. In other ways, though, young Bron was exactly like her rural peers in those salad days, lending a hand in the family business and helping her mother put food on the table. She figures she was probably in high school before she ate meals that her mother hadn't raised or cooked personally. That dietary innocence, however, came to an abrupt end when, distressed by the lack of quality education in their rural community, Weaver's parents opted to send their academically gifted 16-year-old to the famous Webb School in the rural community of Bell Buckle, Tennessee, population just under 400. In yet another unorthodox move, Weaver's farm father accompanied his daughter on her far-flung educational mission, indulging his own lifelong love of the classics by buying out the inventory of a used bookstore and setting up his own shop in Bell Buckle. Out of a class of forty-four, Weaver and two other girls arrived as farmers' daughters of one stripe or another, and they found themselves thriving in their adopted rural town. "I was in a community that regarded intellectual activity very highly. I came from that, and it was part and parcel of my background," Weaver recalls. "I was reading classics, discussing politics, and it was a part of my growing up and growing food and being independent and self-sustaining. . . . I could go toe-to-toe with any of the kids who had come from urban or suburban backgrounds and who were college prepped."

Now, a few decades removed from her high school graduation in 1983, Weaver is glad for the singular agrarian path that propelled her to a degree in geological consulting and an adult life where, with her husband, Bob, she has hung her hat in places ranging from Gettysburg, Pennsylvania, to Antigua, to Naperville, Illinois. The buying of the farm in Geneva, Illinois, in 2003 on the outskirts of Chicagoland, felt less of a leap and more like a "seamless" move back to a life she once knew. "My husband and I decided we were going to recapture a lot of the way I had grown up by creating a farm in which our daughters could be very much involved. . . . It wasn't until a couple of years later that we decided that it would be exciting to turn our love into something that would be a sustainable business."

By 2006 Weaver and her family had incorporated, taking the plunge into a life similar yet different from her upbringing as a Midwest farmer's daughter. Even with Bob's substantial income as a mining engineer, and Bron's profitable work running a geology magazine, they needed the help of a mortgage on their home and 6 acres and the assistance of private investors to afford the dream that would become Heritage Prairie Market. While land in the worn-out hills of Ohio had come cheaply for her parents, prime farm ground in northern Illinois came at a premium—about $26,000 an acre on the edge of Chicagoland before the housing bubble and mortgage crisis dropped the bottom out of the market. It was worth it, Bron and Bob agreed, as a gift to themselves and to their children, to whom they wanted to teach the lessons of growing up rural on a working farm—namely that life has consequences—some that you can control, some you can't.

"That is the most important gift you can give your children, and in farming that's every day," Weaver tells me, the handful of years between now and her decision to return to the land having confirmed in her the truth of those convictions. "That's the weather. That's the consequences of something germinating or not germinating. When you have animals and livestock, there is a dependency and a relationship that is even more visceral than the relationship you have to whatever is growing in your garden. I think that is a tremendous gift that I was given and that my daughters appreciate. I don't know whether or not they'll want to be farmers, or if they'll want to continue the business, but they have an appreciation and a life skill from what they've been exposed to that I think will serve them well." Bron and Bob have four children: Cynthia, serving proudly in the US Army and stationed in Charlottesville with her husband, Seth; Elliot, living in Montana with his wife, Jodi; and high-school-age farm daughters Margaret and Grace. I ask their mom whether she sometimes wonders if the lessons of a true Midwest farmer's daughter will take root in the thoroughly suburban world around her. "I think the farm is what they live and breathe," she says of her teenage daughters. "That's what they have experienced. So I think how they apply it will be up to them."

For Bron and Bob, as for many parents of farm girls, there's an almost subliminal education in the tenderness and nurture woven into the fabric of an agrarian upbringing that transcends parental intention. Weaver likes that the land does its own spiritual teaching. "I can't imagine growing up on a farm and not believing in God. I think that there is that sense that there's something more than what we can easily understand. . . . So many of the kids they know at school don't even have that experience of the impact of nature." After a thoughtful pause, Bron adds, "I want my girls to have that same independence. I don't want them to be one of the herd or to aspire to be one of the herd. They are independent thinkers, and they're independent doers. . . . That heritage could be viewed as a burden by some who wouldn't value it, but I think that they will be better served glorying in their independence as opposed to being burdened by not being like everybody else."

When Bron, Bob, and the girls got serious about running a business on their family farm, they discovered, to their amazement, only one other CSA operating in the area. Initially, they hosted it in the small, on-site market they built to sell Bron's honey, but it wasn't long before they decided they'd rather operate their own CSA. With the purchase of an additional 7 acres, the leasing of 8 more, and the hiring of two farm managers and four staff members, they planted the first seeds in their vision: a place where environmentally aware families could congregate, socialize, and enjoy slow food in a communal atmosphere. "That's what slow food is. Not elitist cocktail hours in the city. . . . And we included the chef community, so that they could help bring some sparkle to the things that we can grow easily here. Building a commercial kitchen and producing food and doing catering for our events was always part of the plan. Because that's the only way that I saw that it would have enough legs that it could support itself," Bron says.

CEO and co-owner, Weaver is serious about running a farm that is something more than an ideological statement. Eschewing the liberalism and preciousness associated with organics in favor of the clear-eyed realism of an Ohio farm daughter, she would have Heritage Prairie Farm operate as a profitable business rather than as a think tank that grows a few veggies on the side to make a political point. She is, she insists, not nonprofit, because a nonprofit couldn't contribute the kind of tax dollars her business enriches the community with, nor would 501c3 status help put her farm girls through college. "There is a lot of pontificating that happens in this industry, but there are not enough people that are doing the hardscrabble business aspect of it," she maintains.

Heritage Prairie Farm is beginning to sell to Whole Foods and to supply microgreens to restaurants like The Big Bowl and Lettuce Entertain You chains in Chicago, in addition to long-established clients like the Marriot on Michigan Avenue. The farm girl in Weaver tells her that to be a reliable player in big-time markets like these, her business needs to operate with the savvy of a Midwest commodity farm, and the socially vested soul of a CSA. It's this combination that she most hopes to teach her crew of four seasonal and two full-time workers. "I take for granted that all the people that work for me are for the most part excited about working with nature. They're excited about growing food, and they embrace organic principles. What I get excited about is teaching this group of young people that there is a livelihood to be made in this business, and that it doesn't mean that you have to forego being able to drive a car." Weaver's all for the industry's feel-good potential, but she'll sleep better at night when affordable,

locally-sourced organics reach the shelves in stores where farmers and factory workers alike can bankroll them rather than merely the urban elites
afforded the right to make emotional or ideological choices at check-out.
"To be able to supply Whole Foods and Dominick's and Jewel with locally
produced food, that's a huge opportunity. There are a lot of livelihoods to
be made here," she insists.

From her post as farm youth mentor-in-chief, Weaver sees firsthand the
trend toward a female-centric organics. Had she broken ground on the farm
in 1985 or 1988, she estimates, 80 percent of her applicants would have
been male. Now those percentages are reversed. Three of her four nonman-
agement positions on the farm are staffed by enthusiastic young women—
not farmers' daughters, but suburban and urban devotees of healthy, environmentally responsible living. Still, sometimes Weaver's experience has
shown her that young women are less inclined than their male counterparts to want to make a true living, rather than a mere lifestyle choice, out
of agriculture. "Men still traditionally gravitate toward a career choice that
will produce a good livelihood to support a family. . . . In the women that
I'm seeing, I don't see the same kind of 'I'll make it a career, and I'm going to be able to support my children and I'm going to be able to buy a car'
mentality. They're not looking at it the same way. Whether it's a freedom
they feel they have, or want—I don't know exactly where all of that comes
from—but I think it goes back to the fact that we haven't demonstrated as
an industry that there truly are livelihoods here."

Among the would-be or wanna-be farmers' daughters that come to
interview at Heritage Prairie Farm, Weaver sees significant commonality.
"They're using this like going back to school . . . as an apprenticeship just
to learn, which is great, but, again, they are not coming here saying, 'In ten
years I want to be your field manager.'" Much of her seasonal crew, Weaver
describes as "women looking for an experience for the season." It's tempt-
ing for Weaver and I, farm kids both, to dismiss such token dirtying of the
hands, still the future of farming belongs as much, if not more, to them as
it does to us. What their generation deems as a farm, or a farm girl, is destined to become its own, no less valuable, agrarian reality.

"The awareness of the connection between health and how you live and
what you eat is something that is becoming more and more mainstream,"

Weaver reminds. "All sectors—suburban, urban—everybody's taking a fresh look at this as an industry." For Weaver it's all part of a cycle, a wheel slowly turning back to its point of origin. "Both of my parents were born in 1921, and then, whether you lived in the city or the suburbs, or whether you lived in a little country town, everybody had a vegetable garden. Everybody everywhere had tomatoes and a row of lettuce in their backyard. I see that swing kind of coming back, and I think that's very, very healthy."

MY INTERVIEW WITH Weaver concluded, I make a detour to Garfield Farm, a 281-acre farmstead and former 1840s teamster inn a few miles west of Bron's place. When Timothy Garfield and his family built the brick inn back in 1846, it quickly became a home away from home for hundreds of teamsters and travelers, a site for jubilant 4th of July dances, a meeting place, and a community watering hole offering good hard cider and better company. Eventually, the coming of the railroad ended the inn-keeping business for the Garfields, though they would continue to till the Kane County, Illinois, soil for nearly a century. In 1977 the last Garfield family owner, Elva Ruth Garfield, made a museum of her family's heritage with a mission to teach others about the prairie life her ancestors had lived.

I find the farm deserted on a Thursday afternoon, the proverbial barn door open wide on the 1842 timber-frame structure. The old inn and tavern where the farmer and his wife kept convivial company is locked up tight and badly in need of tuck-pointing. The only sign of human activity here is a couple of cars in the parking lot and several rare breeds hemmed in behind hand-hewn fences—Java Blacks and Java Whites and some stoical oxen regarding me from behind their stockade. In not so many weeks growers from all over the Midwest will descend on this National Register of Historic Places site to ogle heirloom varieties with names like "Love Lies Bleeding" or "Kiss Me Over the Garden Gate" grown by Thomas Jefferson back when the gentleman farmer-president's vision of an agrarian republic looked like it might indeed come to pass.

Today's Garfield Farm isn't bracing itself against ice storms or grass-hopper plagues, renegade wolves or tuberculosis, but a full-on onslaught of suburban encroachment. Already, it's moved aggressively to ensure that the modest "half-mile view of fields and fencerows" it promises visitors remains unspoiled, slowly raising funds to purchase the 20-acre Mill Creek prairie adjacent to the site. Past the rushing creek looms a housing development tucked behind a man-made lake, the ubiquitous beige vinyl siding-clad structures sporting two-car garages and architectural shingles pleasingly arrayed around the gentle curve of Laura Ingalls Wilder Road.

My day in exurban Kane County, Illinois, leaves me wondering whether agriculture itself is not destined, anywhere within 100 miles of a metropolis, to become agri-tourism, and the modern-day farmer's daughter a re-enactor of sorts, fated to translate a dead language rather than to natively speak it. Perhaps the farmer's daughter is herself a spoil of cultural wars made to embody, like Sitting Bull, a caricature of a proud people passed over if not pacified.

What would a modern-day farmer's daughter wear as her ceremonial garb, one wonders? A gingham dress, perhaps, or a pair of pin-striped Os-hKosh B'gosh overalls? What would she don as her modern war paint—lip balm? What would be her cause worth fighting and dying for—locally grown? What would be her rain dance—the boot scootin' boogie? In the end, many farm relativists contend, so long as men and women, ladies and gentlemen, clamor to see her on her reservation, why shouldn't she oblige her public with dogs and ponies—a real Wild Midwest show? What's wrong with her donning a bonnet and an apron or a tank top and some Levi's, so long as well-meaning folks are willing to pay good money to see her be, for one weekend anyway, what her ancestors were perennially, and without fanfare.

This far into my search for an American icon, however, I am not yet ready to yield to such silo-half-full thinking. Not more than an hour earlier I had been introduced to three farmer's daughters—Bronwyn, Margaret, and Grace—and though Heritage Prairie Farm's handful of acres and staff of six look very different from the Garfield family's historic dominion, maybe in the Google Era, Heritage Prairie is as prairie as any, and heirloom enough for all.

"The American Country Girl. An abundance of sunshine, fresh air, good water, and healthful exercise in the open permit wonderful young life to reach its highest development."

CHAPTER TWELVE

FEMALE FARMERS

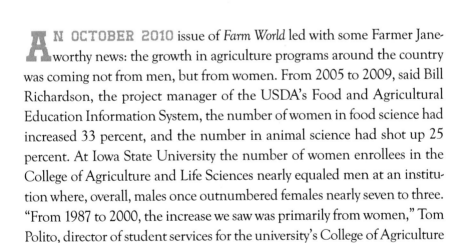

AN OCTOBER 2010 issue of *Farm World* led with some Farmer Jane-worthy news: the growth in agriculture programs around the country was coming not from men, but from women. From 2005 to 2009, said Bill Richardson, the project manager of the USDA's Food and Agricultural Education Information System, the number of women in food science had increased 33 percent, and the number in animal science had shot up 25 percent. At Iowa State University the number of women enrollees in the College of Agriculture and Life Sciences nearly equaled men at an institution where, overall, males once outnumbered females nearly seven to three. "From 1987 to 2000, the increase we saw was primarily from women," Tom Polito, director of student services for the university's College of Agriculture and Life Sciences, told *Farm World's* Michelle Mihaljevich. "They're involved across the board." At Purdue University, the number of female students in the College of Agriculture actually outpaced the number of men.

"Over half of young people I meet now through the Stone Barns Center are women," Fred Kirschenmann tells me when I call to get his reaction to the news. A Distinguished Fellow at the Leopold Center for Sustainable Agriculture and a president at New York's Stone Barns Center for Food and Agriculture, Kirschenmann is blessed with an unusually panoramic view of agriculture. Of the new farmers counted in the most recent Census of Agriculture, 30 percent, he points out, are women.

Kirschenmann is a third-generation farmer as well as a scholar, and his long, thoughtful silences remind of my farming grandfather's. He's farm people, Kirschenmann is, his German grandfather having tilled the Volga River bottomlands in Russia before immigrating to Lincoln, Nebraska, in the late 1800s. His family has been farming since 1930 in Stutsman County, North Dakota, where his parents arrived in time for the Dust Bowl, an event whose lasting trauma still saturates the ethos of his clan. "My dad's near obsession of preventing our land from blowing away was ingrained into me as a child," Kirschenmann writes in an essay published in *Caretakers of Creation*. "As I grew older, he passed on to me his sense of wonder for the miracle of the soil's productivity, as well as a profound sense of responsibility to care for it."

In 2000 Kirschenmann turned over daily operations of his family's 3,500-acre certified organic farm to become the director of the Leopold Center, and, as professor of religion and philosophy at Iowa State, to reflect on agricultural ethics. As president of Kirschenmann Family Farms, he's experienced the highs and lows of rural life with his sister, a Midwest farmer's daughter and decision-making partner. From his vantage he's seen in four generations of Dakotan farmers' daughters (grandmother, mother, sister, and now daughter) how women serve as unheralded determiners of the health and welfare of rural places. His mother worked in the fields every day in the growing season, he tells me, adding, "I still sometimes wonder how she managed to do all the preservation of the food, all of the cooking, and still work out in the field virtually every day while managing to garden at the same time."

Growing up in Windsor, North Dakota, Fred's sister, like many farmers' daughters, left the farm to work in the nearest big city at the age of 19, landing a job at the Bell telephone company and later settling down in Fargo. "My sister was more expected to be a supportive laborer on the farm. . . . I think she to some extent still resents that a little bit. Obviously there was a little sexism involved. Women were not regarded somehow as important in a farming operation as the males were," Kirschenmann recalls. He describes his own daughter, a fourth-generation Midwest farmer's daughter, as an "amazing individual" who left the homestead to earn a degree in dance therapy and now runs her own company helping to answer strategy questions for nonprofits. "She's totally committed to organic agricultural principles," he enthuses of his daughter, an avid gardener, "which she believes is part of the future. She's been extremely supportive in that sense. Often she's given her own time and effort to help think through issues and do feasibility studies and those sorts of things. But she also has never seen herself as being an active manager of the farm." Still, he points out, as she and her partner, both organic advocates, near retirement, they're thinking more and more seriously about getting back into the family business of growing food.

In many ways Kirschenmann's family is a lot like tens of thousands of other long-standing farm families across the Heartland, wherein the latter-day farmer's son often earns his university degree and assumes a managerial or stewardship role while earning his income off-the-farm, while the daughter moves away from agriculture more literally, often deploying farm-born talents in a metropolitan economy. As a result cities around the country, in many of which women significantly outnumber men, abound with farmers' daughters who opted to leave the farm and the prospects of marriage to a farmer's son, but who still hear the call of the land.

Why does Kirschenmann think things have shaken out the way they have for agrarian culture's daughters and sons? "I think there's probably a couple of things that have happened," he observes. "With the feminist movement, and women getting the right to vote, and finally having a place of their own, I think that many of them saw leaving agriculture, and leaving the farm, as part of leaving that past of second-class citizenship. . . . Part of it, too, was to some extent the farmer's daughter culture was a mythology of sort. It was probably nice in literature and movies, but in real life, people didn't always experience it as a positive thing."

When Kirschenmann talks sustainable agriculture in the classroom, where he's inspired many in the College of Agriculture and Life Sciences

at Iowa State University, he often begins by asking how many of his students grew up on a farm: half raise their hands. Next, he asks how many
expect to be involved in the family farm when they graduate, and the percentage drops to a quarter. Finally, he asks how many expect to be "actively
engaged" in farmwork, and the number is reduced to two or three hope-
fuls. "Usually the hands that go up of those who are thinking about going
back to the farm tend not to be women," he notes.

It's not so much that the Midwest's young women at schools of agriculture and technology don't want to be involved in food production and animal husbandry, it's that they're gravitating toward a specific kind of farming—smaller scale, closer to urban centers. Like their male counterparts in
the early nineteenth century, increasingly they're studying agriculture's many
challenges rather than necessary wanting to live them. And when they reach
middle age, many Midwest farm daughters are finding themselves thrust
into a managerial role by dint of inheritance. Kirschenmann reminds, "Over
40 percent of the owners of farms in Iowa are women. Part of that is that
women tend to outlive their husbands, so they end up inheriting the land.
They own the farms and have a lot of say in terms of who they're going to
rent the land to and how it's going to be farmed." And women, the Leopold Center's Distinguished Fellow points out, often have a temperament
that suits them uniquely well to the demands of stewardship—at least that's
been his experience when fielding countless calls from female farm heirs
seeking his input on how best to get the job done.

Kirschenmann resists making a typology of those who seek his counsel. Still, he can't help but notice patterns. "One of the interesting things
I experience at the Leopold Center is that somebody will come to us, and
usually it's a woman in her late 50s or early 60s living on one of the Coasts
. . . and her parents still live on the midwestern farm. They no longer actively farm, but they live on the farm and they have hired a manager. And
they know their parents, who are in their 90s, are not going to be around
forever, so they are starting to think about what they are going to do with
this land, this farm, when they inherit it in the future. They don't want it
be farmed in the way it is now—a typical corn and soybean farm. . . . They
would like to see the land managed in a more sustainable way, and they want
to know where are the resources they can tap into that can enable them to

do that. I have never had a young man come to me with that kind of question. . . . To me, it's a spiritual dimension that drives that." Kirschenmann pauses at this point in his recitation, one of those long, born-on-the-farm pauses wherein one senses still deeper furrows are being plowed. "I don't know whether it's a matter of testosterone or what, but men tend to come at issues from a top-down, controlling, manipulative approach to solving problems, whereas women trend to approach those issues in terms of participation, communication, and working things out. In that sense, I think that women have played a very important role in agriculture and will continue to do so."

While Kirschenmann celebrates the advent of women as food producers and ag managers, he also sees ways in which the encouraging statistics—namely the 30 percent of new farms in the Census run by women—as a likely red herring. "Part of the interesting thing about the latest farm Census data," he explains, "is that of the 300,000 new farmers that are part of that data, none of them are earning more than $10,000 a year. So they are not doing what they're doing to make a living. They're doing it for lifestyle reasons, and yet they are farmers. That's really a 180 degree reversal from where young people were when I grew up."

The question in Kirschenmann's mind is not so much what or who makes a farmer—male or female, Jane or Joe—but a greater concern that the dated 1974 definition by which the Census qualifies a farm (a place that produces $1,000 in gross sales or would have had that place maximized its full production) draws attention away from the rapid rate at which land and capital is consolidating among fewer market players. "I always like to tell people if someone's producing $500 of good food and bringing it to a farmer's market to sell, I want to call them a farmer. . . . So it's not a matter of changing how we come up with the term *farmer*, but that we have to somehow in our stats help the public understand what's actually happening in the erosion of commodity agriculture. As we all know . . . there are huge numbers of farms that are 5,000, 6,000, 10,000 acres. That doesn't mean they're bad farmers or bad farms, but there are some things that we know when you get to that size you can't do."

The statistics Kirschenmann cites both stagger and sober—30 percent of the 2.2 million farms in America earn gross sales of less than $10,000.

Seventy percent of the farms achieved less than $25,000 in gross sales. Seventy-five percent of the nation's total agricultural commodities output is produced by about 190,000 farms—an amount roughly equivalent to the number of farms in Illinois and Iowa alone. In a February 2009 "Editorial Notebook," *New York Times* columnist Verlyn Klinkenborg called the 4,000 new farms created in his home state since 2002 "a genuine source of hope for the American agriculture," citing farmers that were "more diverse than ever, including a high number of women." Still, Klinkenborg's notice, headlined "Good News from Iowa," lamented that those 4,000 new farms were all 9 acres or less, and that "consolidation at the highest level—big farms eating slightly smaller farms"—had quickened.

To hear the United States Department of Agriculture trumpet it, female farmers are all the rage in American agriculture. The USDA cites some compelling data from the 2001 Family Farm Report: in 1978, for example, women owned 5.2 percent of all farms; by 1997, that number increased to 8.6 percent (10.6 percent among nonwhite farmers). Currently, women run approximately 165,000 farms, though that number theoretically includes the many women Kirschenmann flagged who hire a manager to run inherited acres. "Nearly half of these women regard farming as their primary occupation," the USDA website claims, though it leaves out any discussion of how "regarding" one's self as a farmer translates into actual practice. What is harder to argue with is the long-standing fact that women-owned farms are "small, diversified, and financially at-risk," as the USDA describes them. Nearly 70 percent of such farms total 140 acres or less, and nearly 80 percent sell less than $25,000 annually. They are, the data point out, "more likely than other farms to raise livestock or high-value crops." And yet for all the glass-half-full survey data, female farmers and the hope thrust upon them for a radically changed food production system are nothing new.

LIKE HER FRIEND Fred Kirschenmann, Leigh Adcock, former executive director of the Farmer's Union and current head of the Women, Food, and Agriculture Network (WFAN), is the real deal, having grown up on a half-section family farm before pursuing degrees in communications and journalism from the University of Northern Iowa. She's worked in virtually every sector in agriculture, including television, radio, print journalism, public relations, and various environmental and agricultural nonprofit groups, and, as such, has some keen insight into the growing trend of young female agrarians looking to grow food. Adcock herself makes a home on a 5-acre place north of Gilbert, Iowa, where she lives with her husband, Ed, and sons, Richard and David, and where she keeps a flock of laying hens, three goats, and three cats.

"I think a major factor is isolation," Adcock tells me when I ask why more young female food producers have not yet ventured into the open country well beyond the region's college towns and suburbs. "Their major hurdles are access to land, access to capital, and access to relevant information and assistance programs. . . . Local banks in rural areas often are unfamiliar with the small-scale farming model, and are reluctant to lend money, even in the small amounts needed by small-scale farmers. . . . Annual conferences and field days can go a long way toward relieving the feelings of isolation and 'oddness' that many small-scale farmers, especially young women, can face in pursuing their dream of a different kind of agriculture."

Listening to Adcock, as to any skilled advocate, can be beguiling, so much so, in fact, that I find myself forgetting Kirschenmann's sobering read of the real bottom line of the food production system—one where small-scale farmers, however redemptive and welcome their entrance into the market might be, are for the most part economically marginal players. Would yesterday's midwestern farm grandmothers—women who reared and milked and slaughtered, drove tractor, and cooked meals—take exception with today's young female cultivator who grows tomatoes in a small greenhouse and labels herself a farmer?

"Times are changing and agricultural markets are changing with them," Adcock reminds skeptics. "The fastest-growing segment of agriculture in the US is small-scale diversified farming for food production. Organic and natural foods are the largest-growing segments of the ag market. That city gal could make a good living making goat cheese, and that, ultimately, is the bottom line for most 'old farmers.' Whether they find additional value in the health benefits to humans and their environment from small-scale, nonchemical farming is beside the point." Adcock's own father, a traditional Midwest farmer, detested the chemicals required of him to produce a competitive crop under the farm program, Adcock confesses, adding that he would be in favor of any alternatives that would allow a Farmer Jane to make a better, more healthful living.

What has become of the iconic Midwest farmer's daughter, and what goodness might be disappearing along with her, even with the advent of a new generation of Farmer Janes? Adcock's answer gets at a similar disconnect between study and practice that Kirschenmann had earlier reported from his classes at Iowa State University. "I see a definite schism here," she confirms, conceding that young women enrolled at Big Ten land-grant universities like Purdue or Michigan State or Iowa State often fall into two camps—those who experienced the farm growing up and those who became fascinated with it as a purely academic enterprise. She says the members of the first group, the bona fide farmers' daughters, are more inclined to farm conventionally, and enroll in colleges of agriculture to learn the latest techniques for large-scale commodity farming. Conversely, the real passion for farming and food production, Adcock observes, seems to be bubbling up from women who may not have a farm background, and for whom the biggest obstacle isn't education but lack of capital and access to the land otherwise made available by inheritance. "They are coming up with creative ways around it," she explains, "including urban farming and leasing an acre or two from a farmland owner near a viable market. These women are hungry for information and experience in preserving food, and often, in other heirloom skills such as fiber arts, sewing, quilting."

Together Adcock and Kirschenmann confirm a trend the national media has only begun to comprehend: women all over the US are gathering to teach themselves old-fashioned agrarian handicrafts and skills, often at the knee of older female mentors with a lifetime of experience to share. In some key ways they're out-farming the native farmer's daughter, and therein lies the terrific promise of the new-fangled Farmer Jane.

CHAPTER THIRTEEN

FARMERETTES IN THE FARM CITY

HISTORICALLY, THE IDEA that the trail to America's rural daughters wound its way through the Chicago Loop made intuitive, if not paradoxical, sense. Midwest country girl Jane Addams migrated here, after all, from the tiny farm village of Cedarville, Illinois, in 1891, a decade or so after penning her junior class oration, entitled "Bread Givers," at Rockford Seminary. "But while on the one hand, as young women of the nineteenth century, we gladly claim . . . and proudly assert our independence," a young Addams had declared from the lectern, "on the other hand we still retain the idea of womanhood—the Saxon lady whose mission it was to give bread unto her household. So we have planned to be bread-givers throughout our lives; believing in labor alone is happiness, and that the only true and honorable life is one filled with good works and honest toil."

After many months tracing both the routes and roots of Midwest rural daughters like Addams, I have come to the city once-upon-a-time kid milk wagon driver Carl Sandburg dubbed "hog butcherer for the world, tool maker, stacker of wheat," a veritable hive whose "wicked" and "crooked" ways famously lured farm children to their doom or to their destiny. "And they tell me you are brutal," Sandburg wrote in his 1916 poem "Chicago," "and my reply is: On the faces of women and children I have seen the marks of wanton hunger."

In the bustling cities of the Jazz Age Midwest, eager-to-blend immigrants of all kinds shed indigenous identities for the sake of cultural assimilation. As the melting pot heated up in 1915, Chicago's own Martha Foote Crow reassured despairing farm girls that a rural upbringing, far from condemning them to mediocrity, might actually springboard them to eminence. An estimated 80 percent of the names found in *Who's Who in America*, she insisted, had been reared as rural girls or boys. In fact, she reported, the "most valued names in philanthropy and literature" had emerged from the country rank and file, principal among them that "wise helper of all who suffer unjust conditions in city life, Jane Addams." In 1924's *The Woman on the Farm*, Mary Meek Atkeson echoed Crow, describing the farm-raised woman, even if her calling did take her to the city, as "the real lady—the 'loaf-giver'" who pledged herself to "partnership with nature in real creative power."

Circa America's entry into the World War I in 1917, Chicago was already well on its way to becoming the nation's greatest farm commodity city, "the western outpost of a metropolitan economy centered in the great cities of Europe" as historian William Cronon puts it in *Nature's Metropolis*. "Transmutation" was the foundational idea of the great agrarian city, he observes. "Whether one turned dried apples into nails, or salted hams into lumber, or bushels of wheat into bolts of printed cotton, the net effect was to link West with East, rural with urban, farm with factory." The idea worked on a human scale, too, as cities in the hinterland began to supply their most precious human commodity—their farm daughters and sons—to the threshing floor of the city. "To make the best provisions for the transmission of produce is their office," Margaret Fuller wrote of the industrious people of Chicago in 1843, "and the people who live there are such as are suited for this; active, complaisant, inventive, business people. There are no provisions for the student or idler."

From the very beginning the great agropolis astraddle the Chicago River sought symbols befitting its uniquely agrarian contribution, and more often than not those symbols relied on gender for their meaning. In 1885, six short years before Addams began her settlement house on Halsted Street, an unknown artist sculpted the allegoric figures of "Industry" and "Agriculture" as two 5 and a half ton, granite-faced women in the plaza of what would become the Chicago Board of Trade building; later, John Bradley Storrs would cement the ag city's reputation for gendered iconography by capping the building with a monolithic art deco statue of Ceres, the Roman goddess of grain. Even Addams found the need to feminize the city's agricultural genius, her writings consistently tying women's gift for food and fiber to the myth of Demeter and Persephone, and to ancient goddesses she termed, citing anthropologist James Frazer, the Corn Mother and the Corn Maiden. "These deities are always feminine, as is perhaps natural from the association with fecundity and growth," she observed.

In 1918 the farm city hard on the shores of Lake Michigan came to the fore as ground zero in a debate not just about women farmers, "farmerettes" as they were then known, but about whom ought to qualify as a daughter of agriculture. The trigger proved to be a wartime visit to the Windy City by Helen Taft, daughter of former president William Howard Taft, and Virginia Gildersleeve, dean of Barnard College of Columbia University and head of the New York City Women's Agriculture Committee. Upping the irony of the event was Gildersleeve's earlier confession that she didn't know the difference between a carrot and turnip. The *Tribune* article, headlined "Taft's Daughter to Enlist Army of Farmerettes," announced that the meeting would be open to "all women interested in patriotic work" and would discuss "women in farm work." The article further dished that among the topics of discussion would be "plans for the mobilization of a woman's land army to be recruited from all over the United States."

The news of an Illinois women's land army, especially one made up principally of college girls with no background in agriculture, brought immediate reactions from *Tribune* readers. On March 2, the newspaper ran two letters apropos to the controversy, the first written by a reader calling herself "Farmer Girl" and titled "Wanted: Farmerettes." "After reading so many articles on the sorry plight of the farmers, I am wondering why cannot women partially supply the demand for farm labor. Myself a farmer's daughter, I know that most farm women work in the fields during the rush season and generally know as much about the stock as the farmer. When at home I find the farm work more agreeable than domestic service and more healthful." The second letter, written by one-time farm boy S. L. Rapp, cautioned farm-removed Chicagoans from judging the rural folk whose ranks they had only recently left. Titled "Foolish City Boys," it read, "I was raised on the farm and worked there until 18 years old. The remainder of my thirty-one years has been spent in and around cities, and it grieves me to hear our city boys try to discredit their country cousins. I do not wish to discourage the city boy from trying his luck on the farm, but facts are facts. Go to the farm, and after you have mastered the many hard duties of the farmer, you will understand."

Fanning the flames of the developing controversy in America's agropolis was a lengthy commentary published eight days later by "Mme X" in anticipation of yet another, higher profile visit to advocate for a land army of America by England's Helen Fraser. Stories of women's agricultural camps filtering back to Chicago from the East sounded to the approving Madame X "like a delightful modern fairy tale . . . of city girls . . . out for a vacation and yet wanting at the same time to earn their bread and butter." All seventy of the girls who had taken up the plow in Westchester County, New York, the hometown columnist claimed, "were sublimely ignorant of even the rudiments of farming," a fact that caused the inimitable Madame X to wonder aloud, "Once we women get used to the ease and joy of a skirtless condition, do you suppose we'll ever return to bondage?" Fraser's upcoming speech, she assured her readers, would be a can't-miss event for Windy City women seeking "good work, good play, good living, good pay, and good health." Besides, Madame X promised, an "entertaining feature" of the meeting would be "the young and pretty ushers, girls in farmerette costumes."

In its March 15 edition the *Tribune* reported the results of the curious meeting, running a lengthy article under the headline "English Woman Calls Sisters in US Back to Soil." Fraser, the article intimated, had addressed a "large audience of women" and a "sprinkling of men" on the subject of the 285,000 English women working on British farms as part of the war effort of 1918, most of them of the "educated class." "It was difficult at first to convince the farmers that the women would do," Fraser told the crowd assembled that day at the Art Institute of Chicago, "but now they prefer the work of intelligent women to that of boys, which was the only other alternative." The lengthy article was followed by a shorter and less conspicuous item, announcing the no less important but seemingly less sexy call by the State Council of Defense on Adams Street for "men willing to work on farms."

The rural Midwest found itself in unplowed territory in the spring of 1918. Shortages of hired help on large Corn Belt farms were nothing new, but there had always been a pool of free familial labor to turn to—farmers' daughters, sons, and spouses. Now, owing to the mass migration of young rural women and men into the city in the early 1900s, rural America faced what for many of its prideful residents was an unimaginable circumstance—dependence on college-educated city women for relief.

Margaret Day Blake knew she had an idea with legs in the notion of a midwestern women's land army. The college-girl farmerettes Blake recruited aimed to uplift the hard-pressed farm daughter and wife. But she knew, too, that rural women would bristle if do-gooder city gals arrived in country kitchens lacking, as the rural saying goes, "the sense they were born with." Blake's trepidation was based on at least implicit acknowledgment that Midwest farm women had created their own culture separately evolved from their city sisters. Ultimately, this realization prompted her decision that her farmerettes-in-training would not be shipped out to live with women native to the farm, but would instead train sequestered in army fashion, and in that way, render, she hoped, a separate but equal service to the war cause.

A month after Fraser paid visit to Chi-Town, *Tribune* advice columnist Antoinette Donnelly took up the cause of the urban farmerette using what would soon become a familiar argument. "It is not for mere prettiness of face or figure I urge you women to take up the hoe," Donnelly exhorted her readership, "but something bigger, broader, and better." Still, the appeal that followed was mostly a superficial one, as Donnelly promised Chicago's more refined young women the prospect of improved bodies and more robust dispositions. Donnelly quoted the testimony of the mother of a girl who had seen the agrarian light, and who marveled, "Before the end of the season my delicate girl had a color, strength, and an erectness of figure." Elsewhere, while encouraging metropolitan women to "get behind a solider with a hoe in your hand," Donnelly helpfully advocated hand creams to aid in the removal of unsightly soil stains, along with advice for how a garden stood to help "fat" girls who could thereby "bury [their] burdensome adipose." Best of all, becoming a farmerette, the advice-giver intimated, opened up a whole new realm of exciting fashion possibilities. "Wear trouserettes if you will. A short skirt if you won't," she urged, "for the play part of gardening is that unhampered-by-skirts feeling." In conclusion, she maintained, it was time for Chicago's youthful femmes to "face about right and dig."

On May 8 the pro-farmerette *Tribune* upped the ante, running a large photo of "five girls of Bohemian ancestry, stenographers all," who "pulled the covers over their typewriters, put away their pads and pencils, and went home to pack their suitcases" to become farmerettes in Paw Paw, Michigan. The article quoted a letter penned by farm wife Elizabeth Humphrey to the United States Employment Service, wherein Humphrey implored: "Without help, with the breakfast to serve, the kiddies to be made ready for school, then out in the garden and the field, and back to get dinner, you can understand that the girls will be a godsend to me." Accompanying the *Tribune* article was a carefully staged shot of the five young women, dressed to the nines and holding leather valises, standing in what looked to be the door of their outbound train. Predictably, clothiers attempted to cash in on the back-to-the land craze in their summer clothing ads. Featured by Chicago's Chas. A. Stevens and Brothers in the summer of 1918 were the usual silk tricolettes, crepe de chines, and georgettes from $25 to $60. The hottest entry in the summer lineup was a silk gingham dress by which the trendy urbanite could dress the part of the country girl, provided she was willing to pay $20 for the privilege.

By July the buzz created by the *Tribune* and pro-farmerette society ladies Blake and Gildersleeve had mobilized an Illinois division of the Women's Land Army that so impressed one sympathetic husband he offered the rent-free use of his 200-acre dairy farm in Libertyville, Illinois, forty miles north of Chicago, as a working laboratory and training ground. Jumping on the bandwagon were sponsors such as International Harvester, John Deere, and Sears and Roebuck, each eager to capitalize on the publicity the seventy-three farmerettes selected for the program seemed sure to generate. Seventy-five percent of the Farmer Janes, reports Elaine Weiss in her fine book *Fruits of Victory*, were college-educated, and the curriculum at the Libertyville farm closely mimicked the coursework offered at regional agricultural colleges, only on a more compressed timetable. By the end of the summer, Weiss notes, the college girls—a quarter of whom were drawn from the University of Chicago—had coaxed the farm's 187 chickens into laying eggs sufficient to supply the upper crust of Chicago's North Shore, in addition to selling the Libertyville farm's butter, cheese, and green tomato mincemeat elsewhere in city stores.

While the gender-liberal *Tribune* portrayed the urban farms and farmerettes positively, coverage in rural Illinois proved more sanguine regarding the idea of city girls taking their health cure on Midwest farms under the guise of rural relief work. In its August 28, 1920, issue, the *Daily Free Press* in Carbondale, Illinois, ran a front-page item under the heading "A farmerette in real life," describing what they felt to be a true Midwest farmer's daughter, Beulah Boring, aged 10, of Greensburg, Indiana, who had helped her father solve the labor shortage simply by donning her straw hat and bib overalls and heading out to the fields. "She's not a farmerette in a picture," the article noted, twisting the needle, "but one at work." The young Boring, the paper pointed out, held the record for rolling the most acres of corn in one day—21—a good day's work for a man. Other rural small-town newspapers, including the *Freeport Journal-Standard* in Jane Addams's home county, landed somewhere between prideful dismissal of the agriculture suffragists and the offering of back-door laurels. Reprinting a witticism from the *Washington Evening Star*, the *Journal-Standard* of October 5, 1918, quoted a man who commented, "You seem to think well of this farmerette idea," to which an allegorical figure named Farmer Corntossel replied, "Farmerettes save a lot of time by not stoppin' work to swap tobacco and funny stories."

In the end the Midwest farmerette experiment at Libertyville proved productive on a small scale, though the farm itself had mostly fizzled by January of 1919, when Blake announced her midwestern unit would officially withdraw from the national Women's Land Army. After the war a handful of the Farmer Janes went on to either continued university studies in agriculture, to jobs on working farms, or to professional positions with ag companies like International Harvester that had sponsored the short-lived farm in the first instance. In September of 1918 Illinois Governor Frank Lowden visited Libertyville, lending his approval to the straightness of the rows plowed by Jane Brown, a one-time stenographer in a Loop law office turned fleeting female yeoman. The farmerettes likewise impressed *Tribune* reporter Henry M. Hyde, prompting him to observe, "The girls, fresh from the city life, are just as adaptable as their brothers, as they have abundantly shown." Still, the program's zealous public relations campaign almost certainly yielded more than its actual effects on the ground, where, by December 1918, the *Tribune* reported America would be an estimated one million men shy of the number of men needed for planting, harvesting, and cultivating the 1919 crop. The success of the sixty urban farmerettes who stuck with the program through its first harvest seemed somewhat diminished, too, by the rate of attrition, as the paper reported fifteen of the carefully screened young women—fully 25 percent—had been declared "unfitted for work" after the initial two-week trial period.

Back in Chicago and around the Midwest, where the bulk of the women of the Libertyville experiment returned to college or to workaday jobs, the farmerette became a postwar cultural phenomenon. Fundraisers across Illinois in 1922 awarded prizes for best farmerette costume, including the Forest Park Policemen's Benevolent Association Masquerade, where a lucky Farmer Jane impersonator won 5 pounds of bulk coffee donated by the A. Malone Brothers. In Freeport, Illinois, the county seat where Jane Addams's father once mingled easily and well with area corn farmers, the farmerette had entered the costume ball pantheon alongside the "Merry Widow," the "Gibson Girl," the "Tacky Girl," and the "Frivolous Girl." With the boys in uniform reacclimating to civilian life all across the Heartland, the farmerette seemed suddenly the stuff of fairy tale.

AFTER A QUICK detour by the Chicago Board of Trade to witness a flurry of ticker tape decorate the trading pits at the closing bell, I set out for the DePaul Center, where Urban Habitat Chicago's wonder woman, Lee Bouchard, has invited me to a meeting of the Chicago Food

Policy Advisory Council (CFPAC). "If you are an urban agriculture practitioner, concerned community member, or just interested in learning more, we strongly encourage you to attend!" Growing Power's Kelly Trace had written on a listserv announcement heralding the important quarterly meeting.

I arrive to a room full of urban ag enthusiasts and a prompt already written on the dry-erase board: "How does urban ag benefit Chicago?" Long laundry lists of imagined balms have been generated and taped on the columns in the windowless basement meeting space: "Jobs, training, reduce food miles, reduce food deserts, better food, better health, green space, confidence comes from operating, ownership of food and nutrition, allows us to eat food in season, sustainable food sources, teaches people about the environment, tool for addressing different oppressions and dismantling them, teaches people where food comes from, beautification, lower crime."

Chicago has pinned a cornucopia of hopes on its community gardens, which number in the hundreds, as well as on its roughly dozen urban farms. With an estimated fourteen thousand vacant lots citywide, many on the economically challenged South Side, Chi-town appears ripe for a return to a smaller scale, more sustainable agriculture consistent with the municipal housekeeping philosophies of Jane Addams, who once wrote that the "fostering mothers" of the world had, in their desire to grow food for children, "led to a fixed abode and to the beginning of a home." From these seeds, Addams argued, sprung the nation's enduring "domestic morality and customs."

On the front lines of the current fight for the right to make Chicago a farm city is Erika Allen, herself a Midwest farm girl. Allen grew up on the outskirts of a Milwaukee, Wisconsin, farm begun by her dad, former high school all-American basketball player and European professional Will Allen, who in 1993 founded an organization called Growing Power on the north side of Milwaukee near the Park Lawn housing project. Allen's complex of fourteen greenhouses on 2 acres constitute the only land within the Milwaukee city limits zoned as farmland—acres of good dirt on which young Erika learned to grow food as had her father and grandparents.

Allen's grandfolks had been proud to purchase a small farm in Rockville, Maryland, after enduring the indignities of South Carolina sharecropping. In a 2009 interview with *Yes* magazine's Roger Bybee, Allen's father recalled his farm upbringing, saying, "My parents were the biggest influence on my life. We didn't have a TV, and we relied on a wood stove, but we were known as the 'food family' because we had so much food. We could feed 30 people for supper." As the product of a multigenerational African American farm family, who, like many a Midwest farmer's daughter, initially resisted the family calling, Erika moved from Wisconsin to Chicago after high school to pursue a degree in art therapy. But when the chance came to work as the national outreach manager for the organization her father founded, she jumped at it, seeing Growing Power as a way to food-empower at-risk urban populations across the region rather than feeding them through Farm Bill entitlements.

To today's think tank Allen has brought with her a half dozen Growing Power workers, most of them teenage boys who, from their mumbled introductions, appeared equal parts sheepish and jazzed to attend. It is because of Allen's leadership as co-chair of CFPAC, in fact, that we're gathered, discussing a proposed ag farming zoning ordinance. Allen herself has lobbied the city for nearly ten years for zoning laws that would, if invoked, make certain plots within an already pricey city more affordable for would-be urban farmers. In the past Allen has called food "the next frontier of the civil rights movement," and at today's meeting, this Midwest farmer's daughter addresses us both as a practicing farmer and a concerned citizen worried that the agricultural zoning amendment proposed by the city would limit the amount of produce Growing Power could sell at their Cabrini Green food stand and others. "As practitioners, those of us who are urban farmers and actually do the work. . . . We want to protect our communities from a large outside entity coming in and buying up cheap land and putting in a factory-type farm . . . but how do you create stimulation for the community, while at the same time making sure those entities are partners rather than the ones in control of the operations?" she asks a room packed with activists and onlookers. "We want this [urban ag] in our communities," Allen pledges, "and not just in food desert communities; in all communities. We are a diverse group of folks who have different interests, from developers, to folks who are concerned with human rights. To have that broad base of support and to hold folks accountable from our end . . . that makes government do what it's supposed to do."

Grassroots initiatives like Allen's Growing Power have been a fixture in Chicago food politics since indefatigable Midwest Mayor Hazen Pingree popularized the "Pingree Potato Patch" in his home city of Detroit, turning to the city's women to help put his novel plan into practice. In 1893 Pingree and other regional mayors faced conditions not unlike those facing the Corn and Rust Belts in 2011 and 2012—currency instability and credit worries, a badly burst real estate bubble, and an unemployment rate at or over 10 percent. Pingree's novel solution—to enlist church and community leaders to raise money for the seeds and garden tools that could put the municipality's hungry to work in self-cultivated potato patches—caught on around the country, and especially so in nearby Chicago, where the May 20, 1895, edition of the *Tribune* cited with envy the $14,000 worth of potatoes, beans, cabbage, and other vegetables Detroit's garden-sized farms had thus far produced for famine relief.

Much of the success of the potato plots in Detroit could be chalked up to the city's working women, particularly those of eastern European descent, who resurrected and reapplied agrarian survival skills used in the Old Country. Still, all across the Midwest, urban women, including many who were raised on the farm, adopted Pingree's idea with zeal. The April 1, 1895, edition of the *Chicago Tribune* offered two datelined stories concerning the potato patch craze, the first from St. Louis, where an article titled "St. Louis Women Are Interested" read: "The women of St. Louis are coming forward in behalf of the Detroit potato farm plan, and it will not lack henceforth for the encouragement of feminine enthusiasm." A second article assured readers in Chicago that urban ag enthusiasts were on board in Omaha, too, though the scheme was rumored to have failed in Duluth, Minnesota. In the Windy City, ministers such as Reverend Reizel at the North Side Unity Church gathered together to preach the promising future of their farm city. "Acres upon acres of good soil lie idle in and around Chicago surrounded by needy, unemployed people. . . . Here is the open door for the most practical and helpful benevolence; it will yield a large food supply, modifying the distinction between producer and consumer, and make extensive demonstration of public thrift and economy," the reverend told his flock.

Over a century later the group of six women and two men sitting at my table have been asked to document with oversized Crayola markers and butcher paper our vision of what urban ag could do for present-day Chicago. Ours is an eclectic bunch. Two female students from DePaul—Amy and Alicia—have come to the working meeting out of interest stirred by a 400-level class in community development; they sense robust potential for school gardens, garden therapy, and tax incentives for balcony farming. Alicia's parents, she tells us, live in Omaha but own farmland in western Nebraska. She admits that she can feel the tension whenever she tells her traditional farm-state parents what she's learning in her classes about urban ag. "I'm interested to know where people find the threat in urban agriculture, and how to address that threat," she confesses to us. "I wouldn't pit the two groups against one another other, but . . ."

Erica Hougland of Growing Power is next to share what brought her to the front lines of a movement increasingly pitched as a battle. "We focus a lot on anti-racism," she says of her organizational role in Growing Power's Growing Food and Justice for All Initiative (GFJI), which considers how urban agriculture might be a tool for what she calls "dismantling oppressions." Hougland explains, "It's different ways in which people of color communities are usually the communities with food deserts. A lot of the mainstream scholarship is coming from white communities, and we're really trying to break that down and do systematic justice." A friend of Erica's from Chicago's 48th Ward, Joanna, has tagged along, and she too, though a self-professed "novice," finds urban agriculture to be a cause big enough to bring together a community. "I'm working with some other folks on organizing a queer youth community center in Chicago," she informs us, mentioning that she'd like to make an urban garden its centerpiece.

Janice Hill, the farmer's granddaughter and executive planner of Kane County, Illinois, whom I had met several months prior, listens patiently as our group's young women attempt to tease out the origins of the supposed threat traditional ag feels from city farmers. Janice and I are the only participants in our group, or in the groups nearest us, whose parents or grandparents farmed. "I think there's a generational issue oftentimes, too," Hill gently points out. "The average age of a US farmer for a conventional farm is mid- to late 50s. You look around the room here, and though there's a diverse group, you see a young audience. There are also generational issues in terms of passing the land from one generation to the next. . . . A lot of people from rural areas, young people, don't want to live in rural areas. They want to come to the city to farm. And, vice versa, a lot of conventional farmers don't want to move in and deal with a different kind of farming. . . . A lot of people that want to farm don't have land and access to land, and that's something I've been working on, too." Hill unveils to our group a map she has produced in conjunction with the Chicago-based conservation group Openlands of proposed future ag development in her county. She's trying hard, she reaffirms, to keep new development well east of the great swaths of fat farm ground in the less densely populated western side of the county, away from cities like Geneva, Elgin, and Aurora, while identifying unused or underproductive parcels within those inner rings of from 2 to 10 acres that would be "perfect for urban agriculture."

Still, the idea of city-born farmers passing on a true cultivator's knowledge to next-generation Midwest farmers' daughters and sons can seem hopelessly abstract in a highly ideologically environment like today's CFPAC meeting. And yet omnipresent in our discussion are the almost unbelievable events unfolding in Detroit, Michigan, where in the next ten years demolition planners are prepared to tear down an estimated 10,000 residential buildings the city deems dangerous. In sum, 20 to 30 percent of city lots are vacant in the Motor City, and proposals to allow roughly a quarter of the 139-square-mile municipality to revert from urban to semi-rural are being given as serious consideration these days as they were in the era of "Potato Patch" Pingree.

The CFPAC meeting concludes with an around-the-room recitation of each working group's bulleted list of ways to save the city, and the nation, through urban agriculture. At least six full groups deliver action plans wherein Republicans, city officials, and conventional farmers surface again and again as antagonists. A representative of City Farm interrupts his group's youthful reporter to point out that his urban farm generates $130,000 per acre per year, compared to "a typical Illinois corn or soybean farm, which only generates about $1,100 a year per acre." City Farm, he promises, is "one hundred times more productive than a typical Illinois soybean farm."

After the last young woman has covered her last action point—"engaging renters and homeowners in our work and creating sustainable business and program plans and models so that the next generation can continue the work we're doing now"—and after the applause has died down, Janice Hill and I exit the building to linger a moment on the corner of Jackson and State, debriefing on what we've heard. The prospects for a farm city are heady indeed—aquaponics, rooftop gardens, vertical farming—all fruitful potentialities that might allow urbanites to feed themselves and their families, while nurturing a working love of the land in a city-reared child, much as Will Allen's urban farm in Milwaukee did for daughter Erika. In some ways the same grafting of agrarian habits to urban minds is the reason I had the pleasure of speaking with Hill in the first place, the daughter of a Midwest farmer's daughter who married and raised her children with an African American man from Chicago. Hill's mother adamantly refused to move Janice and her two siblings back to rural Michigan, but still she managed to take the fundamentals of an agrarian upbringing with her to the great metropolis.

"In terms of the soil my mom taught us everything," Hill tells me in parting. "We lived the lifestyle of a farm family. We ate like farmers eat . . . my mom cooked from scratch my whole life. We would get a side of beef and put it the deep freeze and fresh vegetables, and although my mom didn't garden, we ate fresh. In that way she was able to take the best part of the farm life and translate it to an urban and suburban setting."

Hills sentiments, I realize, are a fair paraphrase of Jane Addams's, who believed that those who pass down an artisan trade or outlook ought to feel a special obligation to conserve that worldview in the next generation, regardless of where they might live. "One would imagine," she wrote in her 1908 essay "Women's Conscience and Social Amelioration," "that as our grandmothers guarded the health and morals of the young women who spun and wove and sewed in their household, so the women of today would feel equally responsible for the young girls who are doing the same work under changing conditions."

In its way the promise of urban agrarianism in cities like Chicago is the promise of rural-urban synergy, of the vetting and venting of food producers' concerns by entities like CFPAC well-positioned to influence policy changes. In a single afternoon I and fifty or so other farm-vested parties have considered the puzzle of urban agriculture from every angle in the company of more representatives of 501c3s and university departments, centers, and institutes than a midwestern corn-grower could shake a stick at. The number of intellectuals orbiting the urban food movement alone suggests that city minds have indeed turned their attention to food production issues once considered the sole province of the country. While a lack of tillable acres guarantees that the number of Midwest farmers' daughters plying their trade in midwestern metropolises will forever be modest, even that small share brings with it the promise of disproportionate influence, as 60 percent of executive directors of sustainable ag nonprofits are women, according to sustainable farming author Temra Costa. Whether such promising urban leadership is, or ever can be, informed by multiple generations of inherited knowledge of the soil of the kind preserved by generations of Midwest farmers' daughters seems a larger, more vexing question.

CHAPTER FOURTEEN

HER DAUGHTER
HAS A DYNAMO

MANY MOONS—BLANCHE and blue, gibbous and full, harvest and rose—have waxed and waned since first I began my search for the Midwest farmer's daughter. And in that time I have seen more clearly how *what* an American icon chose and *where* she chose it have shaped a nation's destiny.

When the farmer's daughter left her family's green acres in the late nineteenth and early twentieth centuries, she served as the vehicle by which the farm met the city en masse, and ushered in the era of the time clock and paycheck for rural women. As agricultural suffragist and farmerette in World War I, she planted seeds for the back-to-the-land movements that followed; when during World War II she left home to join the Rosie the Riveter at work in bustling East and West Coast shipyards and factories, she commenced the great youth outmigration from aging Midwest farm states that today continues apace. In leaving the farm in the 1960s and 1970s, she and her fellow soil sisters put the lessons of the Sexual Revolution to the test, rejecting long-grounded patriarchies for the egalitarian freedoms of the city. Today the granddaughter of Rosie the Riveter readies herself to begin another back-to-the-land revolution. From the cities she serves as a leader in urban and sustainable agriculture; from the farm, she stands to be a next-generation ag-vocate, consultant, or farm manager.

In 1915 Martha Foote Crow sensed a similar historical flux, one wherein the cultural currency of the country girl seemed to be coming full circle. "The mother of to-day is a bridge between two eras," Crow wrote. "Her mother had a wooden spoon and a skillet; her daughter has a dynamo." The mother of the 1910s, noted Crow, might still claim "that her mother's ways are good enough for her; but the daughter—as between the wooden spoon and the motor, what will she be likely to choose?" Crow contemporary and fellow farm-daughter academic Mary Meek Atkeson likewise envisioned the moment as a fulcrum promising exceptional potential energy. She wrote metaphorically of the peculiar tensions facing the era's farm-vested woman, allegorizing their plight as living between two very different neighbors. "On one side of her may live a family of the old days, with ideas and living conditions little removed from those of the early settlers. . . . On the other side may live a family of wealthy and cultured city people, who maintain a country home for health or pleasure. Yet somehow under these difficult conditions the woman on the farm must evolve . . . not too much like the one nor too much like the other."

In the century following Crow's and Atkeson's distillation of the farm girl esprit de corps, the history of the farmer's daughter eschewed straight lines in favor of poignant circumferences. "Strangely enough, the same influence that took the industrial woman out of the home is to conduct her back again," Crow prophesied. "It will be a regenerated home, one in which the regenerated woman will be able to live." In digital age trends toward do-it-yourself handicrafts, backyard gardens, and work-from-home telecommutes, the prescience of Crow's statement rings truer now than ever. "Every generation," wrote critic Lewis Mumford, "revolts against its father and makes friends with its grandfathers." Surely the same maxim might apply to Google-era agrarian daughters as they weigh the respective cultural and vocational legacies of their mothers and grandmothers.

After years on the trail of an American icon, a journey that took me from the first homestead in Nebraska, to that farm-girl capital of the Dakotas, Fargo, to the heart of the Chicago Loop, and back again to the farm-daughter capital of America, Iowa, a chorus of Midwest farm daughters and granddaughters past, present, and future, still rings in my ears.

For every moment en route I found myself disheartened with my generation's willingness to mourn the farm heritage they forfeited while refusing to vote with their feet to reclaim it, I experienced moments of near certainty that the tide is indeed turning, that a nation's daughters—and sons—have once again, two generations after the back-to-the-land movement, trained their sights on America's rural acres. Six years ago Sara Faivre-Davis gave up a six-figure salary to become a small-scale food producer and mom on Wild Type Ranch, and in her charge to herself rings a possible distillation of a coming zeitgeist: "I woke up one night and thought, 'You know, if you're going to wait until you have all the money you need and you're completely safe and do this for other people's kids and don't do it for your own, then you're a really big hypocrite.'"

Among my many farm-raised interviewees not currently living in the country, a siren song is now heard, louder than ever, calling them back to the greener pastures of their youth. The farmer's daughter nearest and dearest to me, my mother, has, in the years after my father's untimely passing, grown circumspect about her own decision to leave the land, telling me, "Now that I'm older I'd love to live in the country again. And that is my plan to do. I'm going back to my roots."

In Nebraska, Homestead National Monument park ranger Susan Cook still dreams of recapturing in her adult life the child's joy she once felt running horses in the endlessly enchanting Sand Hills: "Every once in a while I think, you know, that is an amazing life because it is so laid back. . . . What's important is important, and *things* don't matter. They *live* life out there."

For her part Kane County Executive Planner Janice Hill has begun the marketing of a homemade documentary, *Deep Roots: Legacies of 150-year-old Family Farms*, and as she does, she asks the what-ifs of continuing her own family's multigenerational legacy on the land. "I always said to my grandmother that I wanted to live down the road from her. There's a pink cottage on her road, and I wanted to live near her when I was younger. I'd say,

'Can I live near you?' . . . You know, if I won a lot of money, I might help to buy our land before we lose it."

Perhaps most hopeful of all are the prospects awaiting Anita Zurbrugg, the Midwest director of American Farmland Trust, whose eldest daughter and son-in-law have returned to DeKalb County, Illinois, and are contemplating a CSA start-up: "We have a 40-acre lot. We could still divide it, so there would be room to build another farming operation. There could be a multigenerational enterprise out there, and I would mostly be the old lady giving unwanted advice."

It seems fitting, then, that I too close up familial circles, ending my search for the Midwest farmer's daughter by returning to where it began, with my own alpha and omega, my great-grandmother Amber. Reading back through the fragments that remain of her childhood letters, I arrive at a passage foreshadowing the choice she would make for a life on the land, a choice that the farm daughters that followed—from my great-aunt to my sister—would have to thank for the many blessings of their respective upbringings. In a letter dated 1912 Amber became indignant with the other boys and girls in her letter circle, writing: "Shame on you boys for not wanting to be farmers. It is the most honest and upright and independent life of all. What would become of the dentists and machinists if it were not for the farmer? Girls, what do you say about this, or are you all against farming as the boys? I am a farmer clear through, and think it is the ideal life to live."

The Midwest farmer's daughter did not become an American icon by indifference or apathy, but via a spirited awakening to her power and calling, real and emphatic. She is not, and never has been, a monolith or a type, yet for her dramatic decline in real numbers over the generations, we feel her spirit enduring still, moving through today's headlines of food price shocks, CSAs, and retro Farmer Janes, vibrant as ever. Certainly, we feel her. Rarely, we see her. Either way, we pay our respects, tip our cap to cheer her, even as she passes.

ACKNOWLEDGMENTS

MANY THANKS TO the farm daughters, farm advocates, and farm experts who kindly agreed to in-person, phone, or e-mail interviews for this project and whose insightful words are included herein. Thank you, Leigh Adcock, Barbara Bardslay, Audra Brown, Daleta Christensen, Susan Cook, Patricia Coon, Angela Crock, David Danbom, Sara Faivre-Davis, Rachel Garst, Beth Grabau, Dale Gruis, Janice Hill, Natasha Jack-Hanlin, Fred Kirschenmann, Becky Kreutner, Laura Krouse, Jamie Leistikow, Gail Logan, Jackie Luckstead, Nicole Patterson, Michele Payn-Knoper, Sarah Ryerson-Meyer, Shaniel Smith, Sarah Uthoff, Bronwyn Weaver, Diane Ott Whealy, Kevin Woods, and Anita Zurbrugg.

Full citation for major, non-interview sources used in the preparation of this manuscript is included in the selected bibliography. Lesser sources, most especially the many newspaper articles read in the preparation of this manuscript, are attributed in-text in accordance with contemporary journalistic practice and per *The Chicago Manual of Style*. Primary electronic databases consulted for this project include the *Chicago Tribune* historical database, *Newspaper Archive*, and the *New York Times* historical database. Historic articles from the *Pioneer Press* and unpublished material from Pickert and Jack family histories are drawn from the author's personal archives and may also be available from the Cedar County (IA) Historical Society.

Thanks to prairie girls Celeste Pille and Lara Blair for permission to reprint their illustrations. "The Prairie Was Her Playground" appears here under copyright of Lara Blair Images (2008), and is featured at Lara's website www.modernprairiegirl.com. Lara Blair is a freelance photographer, photography instructor, and artist who lives with her husband, two girls, and two dogs on an acreage in Washington state. More of Lara's art may be viewed or purchased at etsy.com/shop/larablair. "Farm Life" appears here under copyright of Celeste Pille. Farmer's granddaughter Celeste Pille is an illustrator and graphic artist living in greater Omaha, Nebraska. More of Celeste's graphic art may be found at http://celeste-doodleordie.blogspot.com/.

All historic photos included here and not otherwise attributed appear from *The American Country Girl*, copyright 1915.

SELECTED BIBLIOGRAPHY

"A Breath of City Air." *Confessions of a Farm Wife*. Accessed October 15, 2011. http://webelfamilyfarm.blogspot.com/2011_08_01_archive.html.

Addams, Jane. *The Jane Addams Reader*. Edited by Jean Bethke Elshtain. New York: Basic Books, 2002.

Agnew, Eleanor. *Back from the Land: How Young Americans Went to Nature in the 1970s, and Why They Came Back*. Chicago: Ivan R. Dee, 2004.

Anderson, William. *Laura Ingalls Wilder: The Iowa Story*. Burr Oak, IA: Laura Ingalls Wilder Park and Museum, 1990.

Arngrim, Alison. *Confessions of a Prairie Bitch: How I Survived Nellie Oleson and Learned to Love Being Hated*. New York: It Books, 2010.

Atkeson, Mary Meek. *The Woman on the Farm*. New York: Century, 1924.

Bailey, Liberty Hyde. *Essential Agrarian and Environmental Writings*. Edited by Zachary Michael Jack. Ithaca, NY: Cornell University Press, 2008.

Bartlett, D. W. "The Farmer's Daughter." In *Ladies' Wreath: An Illustrated Annual.* Edited by Sarah Towne Martyn. New York: J. M. Fletcher & Co. 1851, 432-44.

Berry, Wendell. *The Art of the Commonplace: The Agrarian Essays of Wendell Berry.* Edited by Norman Wirzba. Washington, DC: Counterpoint, 2002.

———. *The Unsettling of America: Culture & Agriculture.* San Francisco: Sierra Club Books, 1986.

Bishop, Bill. *The Big Sort: Why the Clustering of Like-Minded America Is Tearing Us Apart.* Boston: Houghton Mifflin, 2008.

Bly, Carol. *Letters from the Country.* New York: Harper & Row, 1981.

Braddon, Mary Elizabeth. "To the Bitter End." *Harper's Bazaar,* May 25, 1872.

Bregendahl, Corry, Carol R. Smith, Tanya Meyer-Dideriksen, Beth Grabau, and Cornelia Flora. *Women, Land, and Legacy: Results from the Listening Sessions.* 2007. Accessed September 30, 2011. http://www.womenland-andlegacy.org/WLL_Listening_Session_Results.pdf.

Brown, Victoria Bissell. *The Education of Jane Addams.* Philadelphia: University of Pennsylvania Press, 2004.

Carpenter, Novella. *Farm City: The Education of an Urban Farmer.* New York: Penguin, 2009.

Coakley, Mary Lewis. *Mister Music Maker Lawrence Welk.* Garden City, NY: Doubleday, 1958.

Costa, Temra. *Farmer Jane: Women Changing the Way We Eat.* Layton, UT: Gibbs Smith, 2010.

Cronon, William. *Nature's Metropolis: Chicago and the Great West.* New York: Norton, 1991.

Crosby, Millard. *She Was Only a Farmer's Daughter: An Old-Fashioned Melodrama in One Act.* New York: Samuel French, 1938.

Crow, Martha Foote. *The American Country Girl.* New York: Frederick A. Stokes Company, 1915.

Danbom, David. *Born in the Country: A History of Rural America.* Baltimore: Johns Hopkins University Press, 1995.

Davidson, Osha Gray. *Broken Heartland: The Rise of America's Rural Ghetto.* New York: Free Press, 1990.

Edlund, Lena. "Sex and the City." *Scandinavian Journal of Economics* 1, no. 107 (2005): 26-44.

Eppes, Allen. *A Maid in Manhattan.* Central Press Association, 1941.

Evans, R. Tripp. *Grant Wood: A Life.* New York: Alfred A. Knopf, 2010.

Fegley, H. Winslow. "The Queen of the Farm and Her Work." *Newcastle (PA) News,* May 21, 1913.

Fellman, Anita Clair. *Little House, Long Shadow: Laura Ingalls Wilder's Impact on American Culture.* Columbia: University of Missouri Press, 2008.

"Financial Position of the Farmer's Daughter." *Iowa Homestead,* October 6, 1898.

Florida, Richard. *Who's Your City?: How the Creative Economy Is Making Where to Live the Most Important Decision of Your Life.* New York: Basic Books, 2008.

Fultz, Jay. *In Search of Donna Reed.* Iowa City, IA: University of Iowa Press, 1998.

Gilbert, Melissa. *Prairie Tale: A Memoir.* New York: Simon Spotlight Entertainment, 2009.

Gilman, Charlotte Perkins. *Herland.* New York: Pantheon Books, 1979.

———. *Women and Economics: A Study of the Economic Relation Between Men and Women as a Factor in Social Evolution.* Boston: Small, Maynard, & Co., 1899.

Gjerde, Jon. "Middleness and the Middle West." In *The American Midwest: Essays on Regional History.* Edited by Andrew R. L. Cayton and Susan E. Gray. Bloomington: Indiana University Press, 2001.

Graham, Nan Wood. *My Brother, Grant Wood.* Iowa City, IA: State Historical Society of Iowa, 1993.

Harkins, Anthony. *Hillbilly: A Cultural History of an American Icon.* New York: Oxford University Press, 2004.

Henderson, Elizabeth, and Robyn Van En. *Sharing the Harvest: A Citizen's Guide to Community Supported Agriculture.* White River Junction, VT: Chelsea Green, 1999.

Howitt, William. "The Farmer's Daughter." *Rural Repository* 20, no. 22 (June 15 1854): 169-72.

Hoyt, Kenneth B. *Future Farmers of America and Career Education.* Washington, DC: US Department of Health, Education, and Welfare, 1978.

Hurt, R. Douglas. *American Agriculture: A Brief History.* West Lafayette, IN: Purdue University Press, 2002.

Iowa State University Extension. *Iowa 4-H Youth Development Participation Policy.* November 2006. Accessed August 22, 2011. http://www.extension.iastate.edu/4h/Documents/Policies4H14ParticipationPolicy.pdf.

Iowa State University Extension 4-H. "Fun and Friends" Mailer. August 2010.

Jack, Walter Thomas. *The Furrow and Us.* Philadelphia: Dorrance, 1946.

Jack, Zachary Michael. *Love of the Land: Essential Farm and Conservation Readings from an American Golden Age, 1880-1920.* Youngstown, NY: Cambria Press, 2006.

Kalish, Mildred Armstrong. *Little Heathens: Hard Times and High Spirits on an Iowa Farm during the Great Depression.* New York: Bantam Books, 2007.

Kimball, Kristin. *The Dirty Life: On Farming, Food, and Love.* New York: Scribner, 2010.

Kingsolver, Barbara. *Animal, Vegetable, Miracle: A Year of Food Life.* New York: HarperCollins, 2007.

Kirbye, J. Edward. "The Creed of Iowa." In *Prairie Gold.* Chicago: Reilly & Britton Co., 1917.

Krouse, Laura. *Final Report for Women Caring for the Land.* May 2009. Accessed September 30, 2011. http://www.wfan.org/Resources_for_Women_Landowners.html.

Lerner, Richard M., Jacqueline V. Lerner, and Erin Phelps. *Waves of the Future: The First Five Years of the 4-H Study of Positive Youth Development.* College Station, TX: AgriLife Communications, 2009. Accessed August 22, 2011. http://www.national4-hheadquarters.gov/library/4-H_Tufts_Report_042009.pdf.

Marquart, Debra. *The Horizontal World: Growing up Wild in the Middle of Nowhere.* New York: Counterpoint, 2006.

Martin, T. T. *The 4-H Club Leader's Handbook: Principles and Procedures*. New York: Harper & Brothers, 1956.

Marty, Gayla. *Memory of Trees: A Daughter's Story of a Family Farm*. Minneapolis: University of Minnesota Press, 2010.

McKeever, William A. *Farm Boys and Girls*. New York: Macmillan, 1912.

New York State Agricultural Society. "The Farmer's Daughter." *Proceedings of the Fifty-first Annual Meeting of the New York State Agricultural Society*. 1892. Accessed October 1, 2011. http://books.google.com/books?id=_7gQAQAAMAAJ&pg=PA442&dq=Proceedings+of+the+Annual+Meeting+of+the+New+York+State+Agrcultural+Society+1892&hl=en&sa=X&ei=4mo1T7e1CJDg2wXQ94jzAQ&ved=0CEIQ6AEwAg#v=onepage&q&f=false.

North Central Regional Center for Rural Development. *Supporting Iowa's Women, Land, and Legacy*. 2007. Accessed September 30, 2011. http://www.womenlandandlegacy.org/WLL-4pager-Nov2007.pdf.

Ott Whealy, Diane. *Gathering: Memoir of a Seed Saver*. Decorah, IA: Seed Savers Exchange, 2011.

Parton, Dolly. *Dolly: My Life and Other Unfinished Business*. New York: HarperCollins, 1994.

Pfaffenberger, Bryan. "A World of Husbands and Mothers: Sex Roles and Their Ideological Context in the Formation of the Farm." In *Sex Roles in Contemporary American Communes*. Edited by Jon Wagner. Bloomington: Indiana University Press, 1982.

Report of the Country Life Commission and Special Message from the President. Spokane, WA: Chamber of Commerce, 1909.

"About Me." *Secret Confessions of an Amish Farmwife*. Accessed October 15, 2011. http://amishfarmwife.blogspot.com/.

Smiley, Jane. *A Thousand Acres*. New York: Knopf, 1991.

University of Illinois Cooperative Extension Service. *4-H News*. November 1978.

———. *Illinois 4-H Members' Handbook*. Urbana, IL, 1959.

Wallace, Henry. *Uncle Henry Wallace: Letters to Farm Families*. Edited by Zachary Michael Jack. West Lafayette, IN: Purdue University Press, 2008.

Waller, Robert James. *The Bridges of Madison County*. New York: Warner Books, 1992.

Weiss, Elaine F. *Fruits of Victory: The Woman's Land Army of America in the Great War*. Washington, DC: Potomac Books, 2008.

White House Rural Council. *Jobs and Economic Security for Rural America*. August 2011.

Wilder, Laura Ingalls. *On the Way Home: A Diary of a Trip from South Dakota to Mansfield, Missouri, in 1894*. New York: Harper & Row, 1962.

Willard, Frances Elizabeth. *Occupations for Women: A Book of Practical Suggestions, for the Material Advancement, the Mental and Physical Development, and the Moral and Spiritual Uplift of Women*. New York: Success Company, 1897.

Worick, Jennifer. *The Prairie Girl's Guide to Life: How to Sew a Sampler Quilt & 49 Other Pioneer Projects for the Modern Girl*. Newton, CT: Taunton Press, 2007.